The Game Needs to Change

This book shows how games can play their part in addressing climate change and how we can sustainably make games for sustainability. Based on both industry practice and research, the chapters within this book give voice to those who are working on sustainability *in* and *through* gaming. From the ways we make games to the ways we design them to resonate with players, these perspectives offer pragmatic recommendations and concrete starting points for working toward sustainability, showing you how to get started, no matter your role.

Divided into four sections, the first section covers systemic approaches to this work, covering how change can occur at all levels of scale. Section 2 explores how stories are told toward sustainable futures across organized communities, emerging markets, journalism, and more. Section 3 critically examines ways of integrating sustainability messaging into games, with practical reflections from developers and researchers on the challenges and strategies unique to this context. Finally, Section 4 highlights individuals who bring theory into action, with personal stories demonstrating what it means to be a game developer dedicated to sustainability.

This book will appeal to game makers who want to address sustainability through the content and message of their creative work, as well as to members of the wider games industry working toward more sustainable practices—including representatives of industry organizations, publishers, platforms, and policymakers. It will also be of interest to game design and development researchers, academics, and students eager to learn more about integrating sustainability into both processes and outcomes.

Patrick Prax holds a PhD in Media Studies and is Professor in Game Design at Uppsala University. In addition to scientific publications, he has presented at GDC, delivered keynotes at games industry conferences, and spoken at the Swedish Democracy Festival. His work focuses on critical game literacy, sustainability, and anti-fascism.

Clayton Whittle is a sustainability educator, serious games designer, and climate psychologist. He received his PhD in Education from Penn State. He was the lead author of the *Environmental Game Design Playbook* and has worked with the Atlantic Council's Climate Resilience Center, Sustainable Forestry Initiative, UN Environmental Programme, and IGDA.

Trevin York has been leading game development teams for over 15 years, making games that bring together engaging entertainment and real-world impacts. His recent work explores how games and play can serve as effective interdisciplinary interventions in complex contexts—like, for example, the climate crisis.

The Game Needs to Change

to Change

Towards Sustainable Game Design

Edited by
Patrick Prax, Clayton Whittle
and Trevin York

CRC Press
Taylor & Francis Group
Boca Raton London New York

CRC Press is an imprint of the
Taylor & Francis Group, an **informa** business

Designed cover image: Sebastian Chaloner

First edition published 2026
by CRC Press
2385 NW Executive Center Drive, Suite 320, Boca Raton FL 33431

and by CRC Press
4 Park Square, Milton Park, Abingdon, Oxon, OX14 4RN

CRC Press is an imprint of Taylor & Francis Group, LLC

Library of Congress Cataloging-in-Publication Data
Names: Prax, Patrick, 1984- editor | Whittle, Clayton, 1984- editor | York, Trevin, 1989- editor
Title: The game needs to change : towards sustainable game design / edited by Patrick Prax, Clayton Whittle, and Trevin York.
Description: First edition. | Boca Raton, FL : CRC Press, 2026. | Includes bibliographical references and index.
Identifiers: LCCN 2025025825 (print) | LCCN 2025025826 (ebook) | ISBN 9781032843667 hardback | ISBN 9781032836195 paperback | ISBN 9781003512400 ebook
Subjects: LCSH: Environmental protection—Computer simulation | Communication in the environmental sciences | Computer-assisted instruction | Video games in education | Video games—Design
Classification: LCC TD170.6 .G36 2026 (print) | LCC TD170.6 (ebook)
LC record available at https://lccn.loc.gov/2025025825
LC ebook record available at https://lccn.loc.gov/2025025826

ISBN: 9781032843667 (hbk)
ISBN: 9781032836195 (pbk)
ISBN: 9781003512400 (ebk)

DOI: 10.1201/9781003512400

Typeset in Times
by codeMantra

Contents

Contributors

Ben Abraham is the research and standard lead for the Sustainable Games Alliance, developing a game industry-specific GHG emissions standard to accelerate climate action. He is the author of the book 'Digital Games After Climate Change', and founder of AfterClimate, which reports on annual game industry emissions. He has a newsletter called Greening the Games Industry.

Sam Alfred is a game designer and programmer. He was the lead designer behind the award-winning ecological strategy game, Terra Nil. He has presented at GDC and been interviewed on NPR. He has an honours degree in Computer Science from the University of Cape Town.

Thorsten Busch is a Senior Lecturer at the Zurich University of Applied Sciences and a Lecturer at the University of St. Gallen, Switzerland. His research and teaching revolve around a wide range of issues in corporate responsibility management, ethics of technology, sustainability, diversity, and game studies.

Florence Chee is an Associate Professor in the School of Communication at Loyola University Chicago, where she directs the Center for Digital Ethics and Policy and founded the Social & Interactive Media Lab Chicago (SIMLab). Her research focuses on the social, cultural, and ethical dimensions of digital technologies, including AI, games, social media, and mobile platforms. She engages across academic, industry, and policy sectors to inform responsible innovation.

Vincenzo De Masi is Associate Professor of Journalism and Communication at Guangdong University of Foreign Studies, China. He received a Ph.D. from the University of Zurich and Lugano. His research focuses on Creative Industries in Asia, social media, metaverse production, and AI video generators. He is the Founding Director of Italian Summer School Veneto and Creative Director of FakeART.

Qinke Di is a teaching assistant at Guangdong Songshan Polytechnic, China. She holds a Bachelor's degree in Cinema and Television from UIC and received a Master's degree from the University of Warwick. Her research interests include digital platforms, labor, and gender, with a focus on female live-streamers' cultural production and labor practices in rural China.

Jennifer Estaris is a Filipina-American game director and designer. She directed the award-winning climate game Monument Valley 3 and Subway Surfers, an endless runner with 4.5B downloads. A climate futurist, she curates eco-punk experimental theatre, co-organizes civic game jams, and speaks globally about the intersection of play and transformation.

In light of the global climate emergency, **Arnaud Fayolle** has shifted his career of veteran Art Director with a track record of 60 released games, including major AAA blockbusters (Forza Horizon series, The Division 2, The Last of Us…) to refocus his efforts on the role of video games as a lever for cultural change in the current socio-environmental crises. He's currently Co-Chair of the IGDA Climate Special Interest Group and works as Sustainability Engagement Director at Ubisoft where he enables developers to reduce their environmental footprint and increase the positive cultural power of their games.

Hexe Fey is an interdisciplinary artist, curator, and interactive fiction writer who works at the edges and confluences of culture, ancestral knowledge, games, movement research and performance, and indigenous futurism. Hexe holds an MFA with a concentration in Indigenous Arts from Goddard College.

Ian Garrett is a designer, producer, educator, and researcher in the field of sustainability in arts and culture. He is Professor of Ecological Design for Performance at York University (Toronto); the director of the Centre for Sustainable Practice in the Arts; producer for Toasterlab, a mixed reality performance collective and media production company; and Venue Manager and Producer for Venue 13 in Edinburgh.

Jonathan Hau-Yoon received a BSc degree in Computational and Applied Mathematics from the University of the Witwatersrand and a Diploma in Visual Communication from the Open Window. He has been a technical artist and game developer for over a decade.

His past work includes contributions to Fortnite, Broforce, GORN, and Genital Jousting, while more recent works lean toward compassion, sustainability, and non-violence.

eileen mary holowka (they/she) is a writer, game dev, and community organizer/advocate with a background in theatre, games, poetry, and patient advocacy. They support indie game developers in running cooperative studios through their work at Baby Ghosts, Weird Ghosts, and Gamma Space Collaborative Studio.

Siyi Li received a Master's degree in Media Communication from Hong Kong Baptist University. She specializes in social media and the metaverse and has authored several articles on how virtual environments shape communication and interaction.

Valley Lopez is an environmental advocate, researcher, and lifelong gamer. His work centers on addressing systemic social issues through policy and alliance building. He has coordinated multiple climate-focused events and led clean energy initiatives, all while maintaining a strong interest in the intersection of art and the environment.

David J Lumb is a senior reporter for CNET covering tech, gaming, and culture. David writes not only longform stories and deep dives but also news and reviews. Over the last decade, he has worked for TechRadar, Engadget, Popular Mechanics, NBC Asian America, Increment, Fast Company, and others.

Aric McBay is an award-winning game developer, climate justice organizer, organic farmer, and the author of nine books including climate fiction novels and non-fiction about building more effective social movements.

Ossian Nordgren is a PhD candidate in Human–Computer Interaction at Uppsala University. An anthropologist by training, Ossian studies technologists in the world. As part of the Digital Ecologies lab, he is currently researching the Swedish gaming field through a critical political-ecology lens.

Cindy Poremba is a digital media researcher, gamemaker, and curator. They are an Associate Professor (Digital Futures) at OCAD University (Tkaronto/Toronto, CA) and Co-Director of OCAD's game:play Lab. Their work focuses on renewable creation practices, game art and curation, capture in postmedia, and interactive documentary.

Tanja Sihvonen is Professor of Communication Studies at the University of Vaasa, Finland. She is specialized in digital media, games, and participatory cultures on the internet. Her most recent work considers role-playing games, TikTok videos, AI, and algorithmic agency in social media.

Yuhan Song is a director and producer in the computer graphics industry with a Master's degree from the New York Institute of Technology. She leads a creative team delivering visual solutions and has showcased VR projects, including a Qing Dynasty costume exhibition with the Palace Museum, at events such as New York Tech Week.

Kara Stone is an experimental game designer and Assistant Professor at Alberta University of the Arts. Her latest project, Solar Server, is a solar-powered web server that hosts low-carbon videogames and can be found at www.solarserver.games.

David Harold ten Cate is a PhD Candidate at Queensland University of Technology's Digital Media Research Centre. His research revolves around the interrelations and contradictions of ecological politics and media aesthetics during the climate crisis.

Danielle Unéus is an interdisciplinary game developer and educator. She has taught game design at Uppsala University and led development at Eat Create Sleep, using play to explore systemic challenges, surface hidden structures, and design for inclusion and critical engagement.

Kristofer Vaske received a Master's degree from Skövde University with a focus on game development and cultural heritage. Since 2023 he has been developing games, exploring how to spread awareness of the climate crisis with a particular interest in portraying movements of activism and resistance.

Anja "Nanna" Venter is a video game producer at Free Lives in Cape Town, South Africa. She graduated with a BA in Visual Communication and has had an illustrious career in the visual arts. She holds a PhD in Media Studies and finished her postdoctoral tenure in Applied Design. She is passionate about the processes and systems that support creative practice.

Stefan Werning is an Associate Professor for Digital Media and Game Studies at Utrecht University. He has been a Comparative Media Studies fellow at MIT and has worked at Codemasters and Nintendo of Europe. His research focuses on ecogaming, game (co-)design as science communication and tool studies.

Hanna Wirman is an Associate
Professor and the head of the 'MSc
in Games' study programme at the
Center for Digital Play at the ITU
in Copenhagen. She has studied and
made games for more than 20 years
with an interest in marginal and
critical ways of playing and making
them. Hanna served on the DiGRA
Executive Board for 11 years, last as
DiGRA's President.

It's Dangerous to Go Alone! Take this Book!

1

Patrick Prax, Clayton Whittle, and Trevin York

We want to contribute to a sustainable world. We want to move beyond greenwashing and empty promises and use the power of games to make real change. If you're like us, know that we made this book for you. Based on both industry practice and research, this book will show you how to approach sustainability *in* gaming and *through* gaming. This book will also show you approaches connected to the physical world and toward honest material impacts. You'll be introduced to well-sourced research perspectives that are connected to and inform pragmatic recommendations and concrete starting points for a meaningful change. You'll read real cases and examples that show you how to get started, no matter where and how you make games.

This book begins by establishing what meaningful sustainability work can look like, contrasting this work to green-washed and surface-level designs, which amount to ineffective, PR-friendly design choices. You'll gain an understanding of the climate crisis that goes beyond common industry lore or public debate. We will unlock the social, political, financial, geographic, and ecological systems of the larger climate/environmental crisis.

However, while the challenge of climate change is vast, there is something in this book for every type of game maker. You'll walk away with concrete steps to take, no matter your specialty, be it design, visual, technical, research, production, or one of countless others. This book will give you a method of connecting these steps to systemic and material goals that respond to the challenge of climate change in a meaningful way. This collection of essays and research uses examples from the industry throughout and gives voice to those who are already doing the good work. This book will show how to use games and their messaging for meaningful change and will discuss how we can improve the way in which we make games while also showing that both of these efforts are necessary in parallel.

This book is by no means the first effort of its kind; the work our chapters build upon is referenced throughout. But we believe this book is valuable because we believe

DOI: 10.1201/9781003512400-1

1

there still remains a long road of critically processing, interrogating, and learning what exactly must be done to get us from where we are to where we want to be.

We will show you that the climate crisis is like an massively multiplayer online game (MMO) raid final boss that will take organized and diverse players to defeat. We will help you arm the party, give you a map and guidance, and show you the goals and best paths. You will find lanterns and rations so that you can make your way through the mines. We will teach you to resist the shadowy voices that want to lure you off the path with promises of shortcuts and greenwashing. While we can give you evidence-based advice and knowledge, we have not yet slain the dragon. Our hope is that you will form a strong alliance along the way so that we all, together, can accomplish this task and bring us a brighter, more humane, and more sustainable tomorrow.

WHO THIS BOOK IS FOR?

This book aims to inform and empower professionals in all areas of the game industry and game making who want to make a difference. We work with and care about games because we believe that they can bring something good into the world. Games offer experiences, friendships, creativity, and economic stability and can even make the world a better place. In this collection of essays, we aim to show how games can do their part in addressing climate change and how we can make games for and about sustainability while walking the walk, running our studios in accordance with our principles, and coming together to make the industry and the world better.

This focus on making games for sustainability and making games sustainably is also visible in the structure of this book, as it speaks primarily to three groups of people:

1. Game makers who want to address sustainability through the content and message of their creative work.
2. Members of the wider game industry, including representatives of industry organizations, publishers, platforms, and policy makers.
3. Researchers and academics who work and teach in the area.

Each section of this book explains the role the chapters play in the overall story of this book, and each chapter has a key takeaway outlined on the first page so that you can quickly find the most relevant perspectives for your work. The combination of well-sourced research texts, pragmatic design examples, and the connection to ongoing practical work in industry organizations like the International Game Developers Association can both inform your own design, community building, and scholarship.

This book is not explicitly written for direct classroom use, but many of the authors have been explicitly working with sustainability issues in game design education, and the wealth of literature and practical examples will be useful for any aspiring game design student and for faculty who want to address the topic in their education.

This book is meant to empower you to work with it in your own context on sustainability questions. The various areas of game design and development as well as different academic disciplines do not (yet) have standardized or commonly accepted definitions of terms and concepts like *Eco games, Green games, sustainability games, environmental games, etc.* In this book, these terms remain nonstandardized specifically so that you, as a reader, can use what works in your context. Our theoretical approach to sustainability and meaningful change is outlined in the next chapters.

HOW TO READ THIS BOOK?

This book has an introduction with two parts, followed by four sections. The introduction chapters outline the theoretical and conceptual framework as well as the theory of change that are the foundations of this book. They outline what meaningful climate action in games is and how to work for real change. The bulk of this book is split into four different foci.

Section 1 continues to explore the systematic understanding and implications of games and the climate crisis.

The section begins with an examination of what systematic and meaningful change means in pragmatic terms, examining how games are intimately connected to the complex sustainability problems that lie at the heart of the ways we produce culture and society. With this grounding in reality in mind, it then discusses social movement games as a direct connection of games and collective action in the real world. The section ends with a chapter that helps you to define the "Unique Impact Potential" for your game and team.

Section 2 is focused on making games sustainably and how we can change the industry we work in.

The section starts from the perspective of games journalists and journalism that concludes with advice for developers looking to navigate this space for their own games. Next, this section takes a closer look at the "green games movement" and offers a discussion on how to approach the sustainability of digital games. The aspect of creative practice and game making in relation to sustainability is highlighted with an analyzed practical example of how you can foreground sustainability in your design by connecting sustainability and nature to the materiality of your creative work. Going from this zoomed-in view on game making to the most collective and systemic topic, the section concludes by examining what the West and market-driven representative democratic systems could learn from China as an example of a space that requires compliance with sustainability policy.

Section 3 shows practical examples and discusses how environmental messaging can be included in the content of your games.

It starts out with a chapter articulating the risk of falling back to greenwashing in the design of the games and shows how to avoid that while maintaining a critical edge, stressing the importance of thinking systemically through your design choices. This section then offers a chapter with one of the best cases for successful sustainability games in existence so far, Terra Nil[1], written by the developers of the game about their logics, process, and the ways in which they managed to include sustainability into the core of their game. The next chapter shows, based on breaking research results and approachable cases, why and how it is central to focus on community building and creating collectives even in the content of our games. The section ends with a chapter about speculative design techniques that can inform making games that allow players to reflect on complex topics themselves without necessarily offering an answer.

Section 4 brings the strands of game production and content together again and adds the perspective of the human game designer.

It starts with a chapter focused on games related to social movements with an analysis of resistance games, but with an emphasis on the makers of these games and their game design as an act of resistance. This connection of design and personal resistance and activism is also the topic for the next chapter where the game director behind Monument Valley 2[2] and 3[3] shares their approach to activism and game making. The section concludes with the most personal chapter included in this book, with the author sharing personal survival strategies as a human making games and working toward sustainability in a system that does not support that (yet).

ACKNOWLEDGMENTS AND THANKS

In addition to the credited contributors, we would like to thank the dedicated work of the members of the Climate SIG of the IGDA. Special thanks go to Paula Escuadra and Hugo Bille for their leadership in this space, as well as their generous advice and kindness.

Patrick: I would like to thank my wife Ida and my children Felix, Justus, Dante, Amadea, Xenia, and My, both for supporting me in this work and for being my reason to hope for a better future. Ich liebe euch!

I would like to thank my colleagues in the Nordic Alliance for Sustainability in Gaming and all the students that joined for our shared events. We need safe spaces to consider these big questions of systemic change, especially in the face of an existential threat like climate change, and this community has been that for me. Finally, thanks to the leadership of and colleagues at my department of game design at Uppsala, for supporting my work and giving valuable feedback.

Clayton: Thank you to my mother, Lisa Erwin, and my sister, Erin, for opening my eyes to the challenges facing our planet. Thank you to Scarlett, my daughter, for inspiring me to think beyond my own generation. Most of all, thank you to Jodi, whose support has made all of this possible. Your love, dedication, respect for my passion for sustainability, and (most importantly) patience for me have made this and so many other things possible. Your kindness and wisdom inspire and empower me. I shudder to think where I would be without you, my love.

And thank you to the IGDA Climate Special Interest Group, specifically Paula Escuadra, Hugo Bille, Grant Shonkwiler, Arnaud Fayolle, and Trevin York. I brought you a 100-page wall of text, and you turned it into a journey that we all could share in.

Trevin: Many thanks to the many colleagues and friends who have become my community even as I've moved thousands of miles around the world. Special thanks, too, to the team behind the Design for Change program at the University of Edinburgh, who transformed how I approach all the work that I do.

Above all, thanks to Cenardach, for her wisdom, for her wit, and for her love.

NOTES

1 Free Lives. *Terra Nil*. Devolver Digital and Netflix. PC and Nintendo Switch. 2023.
2 Ustwo Games. *Monument Valley 2*. Ustwo Games. PC. 2017.
3 Ustwo Games. *Monument Valley 3*. Netflix Games. PC and Nintendo Switch. 2024.

Introduction Part 1—Toward Systemic Sustainability and Games

2

Patrick Prax

ABSTRACT

We need collective, material, and systemic change in order to meaningfully address the climate crisis. This chapter explains why and discusses how this approach is foundational for this book and the following chapters. For this, this chapter draws on perspectives from the game industry as well as research results, UN policy documents, and personal anecdotes. After reading it, you will have a framework for meaningful environmental action, and you will be able to communicate it to others.

KEY TAKEAWAYS

- A framework for understanding and evaluating sustainability efforts in your studio
- A way to think about systemic sustainability that empowers you to resist greenwashing
- Stories, industry data, and research results that enable you to communicate this information to a wide variety of people in a way that connects

DOI: 10.1201/9781003512400-2

INTRODUCTION

During a Nordic sustainability game jam I helped organize last summer, a participant introduced herself as the founder of an indie studio. Her studio had been working on games with a sustainability focus, and she was excited to deepen her knowledge on the topic. I kicked off the jam with a lecture about the need for systemic approaches to sustainability instead of greenwashing, and in response, this indie founder explained that her studio had gotten funding from Meta. Her mood shifted from excitement about her studio and the jam to something more somber as she explained that yes, her team was aware that their game was part of the broader greenwashing efforts of Meta. The social media corporation, she explained, was funding smaller European game studios to fake a commitment to different areas of sustainability, mostly in order to avoid sustainability legislation from the European Union.[1]

Most centrally, Meta wanted to be able to argue for industry self-regulation instead of legal sustainability requirements for social media and digital platforms, largely so that they could continue their unsustainable practices unimpeded. She clarified that if the giant corporation really had wanted to work for systemic sustainability, then it could have put its enormous weight behind real, material, and systemic means to achieve that. Her excitement about her project visibly dimmed as she explained that their game and studio were part of the calculated greenwashing efforts of a cynical tech corporation. This was tough to see, and it left me wondering how many more game makers are feeling the same way: standing behind their game and its pro-environmental message in isolation, but also realizing that in its systemic context, the game studio is even contributing to the greenwashing of a big corporation. There is just no good solution here: Not taking money from Meta to make a game they were believing in is not good either. This is a rock and a hard place, and from the perspective of a single developer, it can feel as if there is no good way of working toward meaningful sustainability.

And indeed, we in the game industry are not satisfied with our industry's climate change and sustainability efforts. This is one of the outcomes of the *2024 State of the Game Industry,* produced by Game Developer and the Game Developers' Conference (GDC), one of the most central annual conferences for game developers,[2] report where the number of highly critical responders has risen from 10% to 27% within a year, and where now 56% are answering that the their company's efforts are either only slightly or not at all successful (Figure 2.1).

The data that the GDC and Game Developer shared here also reflects that people in the game industry are well aware of the difference in impact that a small studio can make in comparison to the tech corporations that own the platforms we create and play on.

> Indie studios aren't the problem. Every indie could go carbon-negative and it wouldn't change a thing. The real culprits are corporations like Amazon and Microsoft who run massively wasteful data centers.[3]

This is not to say that game makers and people in the game industry have been inactive here, waiting for bigger actors to move. On the contrary, they have gotten to work: The Climate special interest group (hereafter SIG) of the International Game Developers

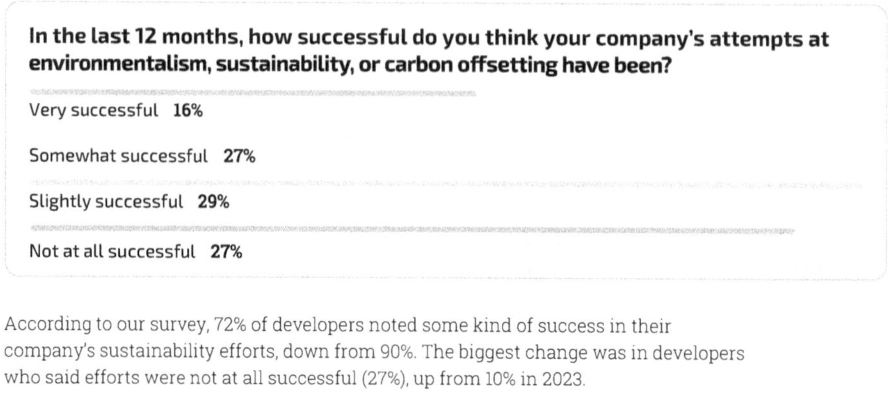

In the last 12 months, how successful do you think your company's attempts at environmentalism, sustainability, or carbon offsetting have been?

Very successful   **16%**

Somewhat successful   **27%**

Slightly successful   **29%**

Not at all successful   **27%**

According to our survey, 72% of developers noted some kind of success in their company's sustainability efforts, down from 90%. The biggest change was in developers who said efforts were not at all successful (27%), up from 10% in 2023.

FIGURE 2.1   Game Developers Conference and Game Developer.com (2023:24). 2024 State of the Game Industry.

Association (hereafter IGDA) is one example. This special interest group has been visible at GDC in presentations and workshops, has published their Environmental Game Design Playbook in 2023,[4] and has been running game jams and regular meetings since 2021. The group has also been connecting with research and teaching on a number of intersections, for example, by inviting relevant researchers into their meetings and having them write about the process[5] in the excellent *Ecogames* book featuring a wealth of academic work on game content and climate.[6] Members have been presenting at academic conferences like the Clash of Realities[7] conference in Cologne in 2022 and been running workshops and panels at GDC.[8]

Other examples of the game industry and academia together tackling sustainability issues include a book about the environmental footprint of the game industry with a focus on the real, material world,[9] the Greening Games Project, which presented its final report in 2023[10] among a number of scientific publications,[11] and Nordic universities have banded together to discuss how to teach about sustainability, resulting in a yearly game jam and exchange that shows students the somber reality and yet still inspires hope and creativity.[12]

The material reality of the game industry is dire. The significant carbon emissions and energy usage of the developers, distribution, and players are only the beginning.[13] Digital games are also made possible by and in turn enable the colonial exploitation of the planet and its people[14] all while normalizing designed obsolescence in the very core of our tech and cultural industries, even though there are recent signs for a move away from this approach. In a later chapter, Busch et al. offer an extensive overview over the various sustainability crises that games and gaming are connected to.

In summary, we need to radically change the ways in which we produce society on a systemic level to address the challenges of climate change. This also means that we need to take a close look at not only what messages we build into our games, but also at how we make these games. Because games and the industries they are a part of, tech and culture, are so intimately connected to the climate crisis, we also need to examine the context in which they are produced. There is already grassroots work being done in building organizations, collecting knowledge, and building connections between research and industry. However, if it is literally possible that entire game studios are the greenwashing of big tech corporations, even while making legitimate sustainability games, then we really need to think outside of the box to reach the kind of holistic and systemic change that we need. In the face of these challenges, this chapter asks:

> "What would constitute successful and meaningful climate action in the games industry?"

SYSTEMIC CHANGE

The report *Making Peace with Nature* from the United Nations Environment Programme explains that the needed "Transformation will involve a fundamental change in the technological, economic, and social organization of society, including world views, norms, values and governance."[15] From this perspective, if there is to be any chance to avoid climate catastrophe, climate change needs to be addressed through real, material changes in our global production systems, mechanisms of governance, and in the way we structure our economic systems and incentives. These changes need to be accompanied by cultural, societal, and normative changes, but those immaterial elements cannot stand alone. In short, we need to radically rethink how we produce society.

Here, the fact that games are, as mentioned above, in one way or another connected to many of the biggest sustainability issues we are currently facing, might also put us in a position to create positive change in many places. Especially digital gaming could be an influential starting point for said material as well as a discursive sea change.

This perspective already exists in the game industry. Taking again the data from the 2024 State of the Game Industry (GDC & Game Developer, 2023), the report chooses to cite a number of very interesting quotes, for example this one:

> They have been diligent but there is still the silent belief that 'first we make money, then we clean after ourselves,' without accepting that this is pretty much the mindset that brought the global society to this state. The environmental crisis seems a much larger problem that my current company believes to be able to tackle.[16]

This shows that there already is an understanding that the climate crisis and sustainability issues need to be addressed system-wide, with actions that go beyond the level of the individual studio. The quote also calls for us to not consider sustainability as an afterthought but to treat it as a priority from the start. As a takeaway here, a sustainable game industry would have to be part and object of a radical and systemic change in how it operates. This change needs to also encompass the material world and needs to extend beyond discourse and discussion.

IMPEDING TRANSFORMATION

However, the UN report also explains that "some attractive and feasible actions can impede transformation,"[17] and transformation here means the fundamental change to how we build society that is mentioned in the paragraph above. As a mechanism for this impediment of transformation, the report explains that "The changes that appear most feasible may be those that do not contribute to, or even impede, transformative change, for instance by retaining or even consolidating the power interests vested in the status quo (see Section 5.3)."[18] Here it says that "Some individuals and organizations also have substantial stakes in maintaining the status quo. These vested interests may oppose changes that disrupt their livelihoods, market shares and future revenues."[19] This is the United Nations stating two things:

1. People and groups in positions of power will obstruct meaningful and systemic climate action because it would impact their wealth.
2. They can distract us from focusing on the necessary systemic change through ineffective, but enticing alternative actions, e.g. greenwashing

Parts of this warning can be applied in a fairly straightforward manner to digital games. Games are published by and played on the platforms of some of the most powerful and influential international corporations on the planet. They, at least in the short term, benefit from the continuation of the hardware race that they are currently winning. Chapter 5 will offer more pragmatic ways to think about systemic change and there we will come back to the responsibility of powerful actors and organizations.

GREEN WASHING AND INDIVIDUALIZATION— WHY WE NEED COLLECTIVE ACTION

What are the "attractive and feasible actions [that] can impede transformation"?[20] One way of understanding this formulation relates to a common narrative about climate change in both contemporary culture and games that presents it as the responsibility of individuals instead of groups and mostly a question of what we believe instead of how we collectively act. In a call for action, preceding the urgent warning of the United Nations by 20 years, Maniates[21] states that "the forces that systematically individualize responsibility for environmental degradation must be challenged." This narrative positions individuals and their actions as the final and sole responsible for climate catastrophe and, importantly, individual behavioral change as the only avenue for change. It disguises material, and systemic problems as exclusively issues of individual virtue and character and makes it impossible to even consider the role of the vested industrial interests that benefit from the status quo, like the fossil fuel industries.[22] This raises the question of whether the individualization of climate responsibility then is not only an involuntary consequence of neoliberalism,[23] but a purposeful distraction and greenwashing by the fossil-fuel industry.[24][25] Even the concept of the "carbon footprint" is a development of the Exxon Mobile concern as a distraction,[26] so that we all focus on the responsibility of each individual instead of coming together to change the world. Again: The carbon footprint, the entire metaphor, is toxic greenwashing from oil companies. This is what the UN is warning us about: the vested interests benefiting from the maintenance of the system that is destroying the planet, and that those might hinder effective and systemic climate action.

Here then, the individualization of climate action can be seen as a kind of purposeful greenwashing: an attempt to create a narrative in which the systemic change of society cannot even be imagined, and so the necessary systemic and radical transformation becomes impossible to even consider. Maniates's[27] call might not have been meant to address games specifically, but it remains relevant while one of the proposed uses of games is the education of players. Research about sustainability education has shown that it is both possible and necessary to go beyond placing the responsibility for climate change at the feet of customers in a market and to address sustainability, together with the students, from a systemic and interdisciplinary perspective.[28][29][30] In addition, games are perfectly suited to portray the workings of systems and to open them up for critical reflection in ways that other media are incapable of.[31][32][33][34] Educating players to see the workings of destructive and exploitative systems and the ways in which they are implicated (and the games they are currently playing are implicated) is definitely a good start and something that we will come back to later. However, the conclusion here for defining a sustainable game industry is that it has to resist individualization and include collective and organized action. This is also something that the industry on some level already understands.

> Others spoke to feelings of disillusionment, frustration, and helplessness when it comes to making a positive impact on the environment—especially related to the issue of personal versus collective responsibility.[35]

BEYOND DISCOURSE TOWARD MATERIAL CHANGE

It is a good step to show, also in the content and message of the games that we make, that we must approach sustainability questions collectively and that it is counterproductive to focus on individual responsibility. This is also typically the first thing we think about when considering games and sustainability. Games as culture can change the way players think and feel about their environment and can in this way create societal change. The possibility of changing the way players see the world and to inspire them toward systemic change is valid. That said, it should not be seen as a reason to keep polluting or being part of the systems that create climate change. It is not only about making games for sustainability, but also about making games sustainably. This is a central point, and one that other authors in this book discuss from different perspectives. David ten Cate discusses the need for ecological games to also consider how they are being made as a part of an economic system and Jennifer Estaris discusses from the view of a designer how she sees the potential of games for her concept of sea change and the connections of what is in our hearts and how we act. So how can we make sure that we together also take that next step, reach beyond discourse, and collectively change our systems?

MATERIAL AND COLLECTIVE CHANGE

At the first planning meeting for the Nordic sustainability-themed game jam, a member of the Finnish game industry presented the work they had done to get their studio to address sustainability issues. Amazingly, and at that point ahead of their time, they included emissions by the players who were playing their game into the carbon emission calculations of their studio. These emissions during play are often one of the biggest sources of emissions related to digital games and considering these the responsibility of the studio shows an honest and real engagement for sustainability. They then bought carbon offsets from the highest-rated sellers that they could find to make up for their emissions. However, this is also where their efforts ended. First, it is important to say that this is not intended as an example of a member of the game industry being naïve or doing something stupid. On the contrary, I would consider this an exemplary effort to do their part in addressing the climate crisis. That being said, it is important to also admit that, in this discursive prison of the individualization of climate responsibility that has been built in contemporary culture for decades, approaches to material change have also been only allowed to be imagined in limited ways. Here, material change starts and also ends with carbon offsets.

Carbon offsets have been heavily criticized for various reasons in research. Offsets frequently do not really work, binding carbon only for short periods but calculated as if the carbon would stay quasi-permanently bound, and with a lack of oversight are open for exploitation, cheating, and the continuation of the same colonial and logics that constitute the climate crisis.[36][37][38] They even continue to perpetuate the mindset and system of trading the right to pollute, along with the mindset of individual customer responsibility that the polluting industries promote.[39] At this point, issues with carbon offsets are so widely known, and in their worst excesses so absurd as to be funny, that they were a topic of the comedy show "Last Week Tonight" in 2022.[40] At the time of writing in late 2024, we're once again in a round of CEOs getting sued because their companies overestimated their emission reductions by an average of 1000%.[41] Here, carbon offsetting is an example of the "attractive and feasible actions" that "can impede transformation."[42]

This shows that even a collective, systemic intervention like carbon trading and the offset industry can be insufficient or even something that impedes transformative change, especially if done instead of reducing energy use or a restructuring of technological infrastructure. The perspectives from the industry already share this conclusion.

> We do some carbon offsetting, but that feels like a Band-Aid on a bleeding wound. Not much is being done, or not that I know of, to address deeper issues like the energy consumption taken up with processor-intensive game development and maintaining our massive internal networks.[43]

This is why it was heartbreaking to see the honest and commendable effort of our Finnish colleague end with carbon offsets.

THINKING IN "SYSTEMIC CHANGE"

At the time when this happened, I was not prepared to offer real alternatives. Honestly, it would still be difficult. This speaks to how effective the individualization discourse is that we have difficulties even thinking about collective changes. Systemic change is tricky because we are used to the systems we live in, and it can be challenging to see a bigger picture. It also often requires some kind of safe space and trust to be able to pitch systemic change, because any suggestion of systemic change can frequently result in having to present and justify a ready and solved alternative system which is of course impossible to do. If the justifications that are demanded from any change and alternative are not also required from the status quo, which is currently destroying our planet to the point that the UN calls for fundamental change, then this indicates a bias toward the existing system that is likely in some way in the head of each of us, and possibly an unsafe space. This book then also tries to both offer some examples for what such change could look like, so that we can all use them in conversations and in our thinking, and it tries to grow the community of game workers who already care about this conversation and who can be this safe space for all of us.

In a later chapter, I will offer more examples for collective, material, and systemic change that are meant to be starting points for finding new ways of meaningful climate action. This is work in progress.

But let us consider this example: Some games already offer an Eco Mode. This means that players can choose to play the game at lower performance to reduce emissions and energy costs. What if we argued, together, for legislation that required games to show the player a summary of their gaming energy use and estimated emissions at the end of each week, kind of like your phone shows you which apps drain your battery the most when it is about to die? Now we would take a first step into a conversation around the environmental responsibility of designers and players and start a competition for creating the most frugal games that get the most amazing performance out of the least amount of emissions. There could be collective achievements for players connected to the use of Eco Mode, and then at a later step, once we have built acceptance for eco modes, they could become industry default and even policy. This would be a longer process, but we have already taken the first steps and with our eyes set to collective, material, and systemic change we can continue this path together.

CONCLUSION

This chapter asked the question: "What would constitute successful and meaningful climate action in the games industry?"

Based on the discussion of industry feedback in connection to research and the reports of the United Nations, successful and meaningful climate action in the game industry would:

1. move from individual to collective responsibility and action,
2. go beyond the discursive toward **material change**,
3. **resist greenwashing** that (purposefully) impedes transformative change, and
4. lead to **systemic change** in the way we produce society.

Returning to the story of the indie developer working on a sustainability game with money from Meta, I would like to argue that many of us can on some level relate to their experience. The results from the survey here also show people in the game industry are on some level aware that a lot of the sustainability work we are doing or are allowed to do can be either diverted by or even a part of greenwashing. This is where this chapter, and collectively this book, argues that we need an approach that centers around **collective action to reach systemic and material goals**. This is what this book is all about.

So, let's get into it!

NOTES

1 Big tech has regrettably been combating EU regulation for a long time (Doctorow, 2024) and instead favors a self-regulation approach. However, "a critical approach is essential when evaluating Big Tech's often deceptive sustainability narratives and underscores the need for more rigorous regulatory frameworks" (Vrikki, 2024:1) in all areas, but specifically around sustainability. "disclosure legislation, international soft law and private actors' corporate sustainability codes of conduct. Despite an abundance of norms, egregious human and environmental rights violations […] continue." (Zumbansen, 2024:1; about sustainability regulations in global value chains)

Cory Doctorow, "Big Tech to EU: 'Drop Dead,'" Electronic Frontier Foundation, May 13, 2024, https://www.eff.org/deeplinks/2024/05/big-tech-eu-drop-dead.

Photini Vrikki, "Measuring Up? The Illusion of Sustainability and the Limits of Big Tech Self-Regulation," *Sustainability* 16, no. 23 (January 2024): 10197, https://doi.org/10.3390/su162310197.

Peer C. Zumbansen, "De-Valuing Sustainability: Financialized Disclosure Governance And Transparency In Modern Slavery And Climate Change," SSRN Scholarly Paper (Rochester, NY: Social Science Research Network, December 13, 2024), https://doi.org/10.2139/ssrn.5054382.

2 Game Developers Conference and GameDeveloper.com (2023:24). 2024 State of the Game Industry.

3 GDC & Game Developer, 2023.

4 Clayton Whittle, Trevin York, Paula Angela Escuadra, Grant Shonkwiler, Hugo Bille, Arnaud Fayolle, Benn McGregor, Shayne Hayes, Felix Knight, Andrew Wills, Alenda Chang and Daniel Fernández Galeote, (2022). *The Environmental Game Design Playbook (Presented by the IGDA Climate Special Interest Group)*. International Game Developers Association.

5 Alenda Y. Chang, "Change for Games: On Sustainable Design Patterns for the (Digital) Future," in *Ecogames*, by Laura Op De Beke et al. (Nieuwe Prinsengracht 89 1018 VR Amsterdam Nederland: Amsterdam University Press, 2023), https://doi.org/10.5117/9789463721196_ch01.

6 Laura Op De Beke, Joost Raessens, Stefan Werning and Gerald Farca, (Eds.). (2024). *Ecogames: Playful Perspectives on the Climate Crisis*. (Nieuwe Prinsengracht 89 1018 VR Amsterdam Nederland: Amsterdam University Press, 2023), https://doi.org/10.5117/9789463721196_ch01.

7 Clah of Realities Conference. Cologne, 2022. https://clashofrealities.com/2022/

8 *Patrick Prax, Clayton Whittle, Trevin York, Sonia Fizek.* (2023) Educators Summit: Teaching Sustainability and Game Design: From the Low-Hanging Fruits to the Root of the Problem. GDC Educators Summit.

https://schedule.gdconf.com/session/educators-summit-teaching-sustainability-and-game-design-from-the-low-hanging-fruits-to-the-root-of-the-problem/891032/?_mc=edit_gdcsf_gdcsf_le_x_41_x_2023

9 Benjamin J. Abraham, *Digital Games After Climate Change*, Palgrave Studies in Media and Environmental Communication (Cham: Springer International Publishing, 2022), https://doi.org/10.1007/978-3-030-91705-0.

10 *Greening Games Education_Report 2023*. Google Docs. Retrieved March 20, 2024, from https://docs.google.com/document/d/1caJyVQ_Tcwzy8l_4bt1Zl3WN2u2hqR3Feb8M_UjGw8s/edit?usp=embed_facebook

11 Sonia Fizek et al., "Teaching Environmentally Conscious Game Design: Lessons and Challenges," *Games: Research and Practice* 1, no. 1 (March 12, 2023): 3:1-3:9, https://doi.org/10.1145/3583058.

12 Hanna Wirman et al., "From 'Doomsday Talks' to 'Excitement at Last': A Case Study of Incorporating Sustainability Perspectives into Formal Game Education," in *Nordic DiGRA Conference 2023, Uppsala, 27–28 April 2023*, 2023, 1–10, https://pure.itu.dk/files/100885374/NoDigra_2023_paper_7722.pdf.

13 Abraham, *Digital Games After Climate Change*, 2022.

14 Patrick Prax, "Are the Bullets Going over Our Head? Designed Ambivalence in the Representation of Armed Conflict in Games," in *Representing Conflicts in Games* (Routledge, 2022), 153–170, https://www.taylorfrancis.com/chapters/edit/10.4324/9781003297406-13/bullets-going-head-designed-ambivalence-representation-armed-conflict-games-patrick-prax.

15 United Nations Environment Programme [UNEP] (2021). Making Peace with Nature. A Scientific Blueprint to Tackle the Climate, Biodiversity and Pollution Emergencies, Nairobi, https://www.unep.org/resources/making-peace-nature. Page 13.

16 GDC & Game Developer, 2023.

17 United Nations Environment Programme, 2021. Page 113.

18 United Nations Environment Programme, 2021. Page 114.

19 United Nations Environment Programme, 2021. Page 104.

20 United Nations Environment Programme, 2021.

21 Michael F. Maniates, "Individualization: Plant a Tree, Buy a Bike, Save the World?," *Global Environmental Politics* 1, no. 3 (2001): 31–52.

22 Maniates, 2001.

23 Jennifer Kent, "Individualized Responsibility and Climate Change:'If Climate Protection Becomes Everyone's Responsibility, Does It End up Being No-One's?'," *Cosmopolitan Civil Societies: An Interdisciplinary Journal* 1, no. 3 (2009): 132–149.

24 Magali A. Delmas and Vanessa Cuerel Burbano, "The Drivers of Greenwashing," *California Management Review* 54, no. 1 (2011): 64–87.

25 William S. Laufer, "Social Accountability and Corporate Greenwashing," *Journal of Business Ethics* 43, no. 3 (2003): 253–261.

26 Geoffrey Supran and Naomi Oreskes, "Rhetoric and Frame Analysis of ExxonMobil's Climate Change Communications," *One Earth* 4, no. 5 (2021): 696–719.

27 Maniates, 2001.

28 Jo-Anne Ferreira, "The Limits of Environmental Educators' Fashioning of 'Individualized' Environmental Citizens," *The Journal of Environmental Education* 50, no. 4–6 (December 2, 2019): 321–331, https://doi.org/10.1080/00958964.2019.1721769

29 Jo-Anne Ferreira, Vicki Keliher, and Jessica Blomfield, "Becoming a Reflective Environmental Educator: Students' Insights on the Benefits of Reflective Practice," *Reflective Practice* 14, no. 3 (June 1, 2013): 368–380, https://doi.org/10.1080/14623943.2013.767233.

30 Elina Eriksson et al., "Addressing Students' Eco-Anxiety When Teaching Sustainability in Higher Education," in *2022 International Conference on ICT for Sustainability (ICT4S)*, 2022, 88–98, https://doi.org/10.1109/ICT4S55073.2022.00020.

31 Ian Bogost, "Persuasive Games, A Decade Later," in *Persuasive Gaming in Context*, edited by Teresa de la Hera, Jeroen Jansz, Joost Raessens, and Ben Schouten (Amsterdam University Press, 2021), 29–40. https://doi.org/10.2307/j.ctv1hw3z1d.5.
32 Michael Mateas, "Procedural Literacy: Educating the New Media Practitioner," *On the Horizon* 13, no. 2 (2005): 101–111.
33 Dennis Meadows, Linda Booth Sweeney, and Gillian Martin Mehers, *The Climate Change Playbook: 22 Systems Thinking Games for More Effective Communication about Climate Change* (Chelsea Green Publishing, 2016), https://books.google.com/books?hl=sv&lr=&id=BxMRDAAAQBAJ&oi=fnd&pg=PR9&dq=games+and+system+thinking&ots=XwsIOqQb7c&sig=Nby_ZhAcvTDmIrIYqJ8MXAEA20Y.
34 Pejman Sajjadi et al., "Promoting Systems Thinking and Pro-Environmental Policy Support through Serious Games," *Frontiers in Environmental Science* 10 (2022): 957204.
35 GDC & Game Developer, 2023.
36 Inés Acosta, "'Green Desert' Monoculture Forests Spreading in Africa and South America," *The Guardian*, September 26, 2011, sec. Environment, https://www.theguardian.com/environment/2011/sep/26/monoculture-forests-africa-south-america.
37 David Bickford et al., "Science Communication for Biodiversity Conservation," *Biological Conservation*, ADVANCING ENVIRONMENTAL CONSERVATION: ESSAYS IN HONOR OF NAVJOT SODHI, 151, no. 1 (July 1, 2012): 74–76, https://doi.org/10.1016/j.biocon.2011.12.016.
38 Daniel J. Sherman, "Upsetting the Offset: The Political Economy of Carbon Markets," *Social Movement Studies* 12, no. 3 (2013): 354–355, https://doi.org/10.1080/14742837.2013.787766.
39 Steffen Böhm, Maria Ceci Misoczky, and Sandra Moog, "Greening Capitalism? A Marxist Critique of Carbon Markets," *Organization Studies* 33, no. 11 (November 2012): 1617–1638, https://doi.org/10.1177/0170840612463326.
40 Adrian Horton, "John Oliver on Corporate 'Net Zero' Proposals: 'We Cannot Offset Our Way out of Climate Change,'" *The Guardian*, August 22, 2022, sec. Television & radio, https://www.theguardian.com/tv-and-radio/2022/aug/22/john-oliver-net-zero-climate-change-last-week-tonight.
41 Patrick Greenfield, "Ex-Carbon Offsetting Boss Charged in New York with Multimillion-Dollar Fraud," *The Guardian*, October 4, 2024, sec. Environment, https://www.theguardian.com/environment/2024/oct/04/ex-carbon-offsetting-boss-kenneth-newcombe-charged-in-new-york-with-multimillion-dollar.
42 United Nations Environment Programme, 2021.
43 GDC & Game Developer, 2023.

Introduction Part 2—Designing with Theories of Change

3

Clayton Whittle and Trevin York

ABSTRACT

"Across the first section of this book, we will show you what systemic and collective change can mean in practice, how to connect games to collective action during design, and why this broad collective approach really is necessary."

KEY TAKEAWAYS

- The difference between empowering action and shifting behavior, and why we're focused on the former
- Common predictors of pro-environmental action and how to begin thinking about designing with these in mind
- Key examples of theories of action put into practice

In the early months of 2010, a teenager in Nairobi named Amina sat in a crowded internet café, eyes fixed on the glowing screen before her. She had just embarked on a mission from a mysterious network known as the Evoke. Her task is to identify a pressing issue in her community and devise a solution. With determination, Amina chose to tackle the problem of food waste in her neighborhood. She documented her observations, proposed a community composting initiative, and shared her plan on the Evoke platform. Encouraged by feedback from peers around the world, she refined her project, transforming a simple idea into a tangible action plan.

DOI: 10.1201/9781003512400-3

This scenario was part of *Urgent Evoke*,[1] an alternate reality game launched by the World Bank Institute in March 2010. Designed by game theorist Jane McGonigal, the game aimed to empower young people, especially across the African continent, to develop innovative solutions to global challenges such as hunger, water access, climate change, and poverty. Over ten weeks, participants engaged with weekly missions presented through graphic novel episodes, each focusing on a different societal issue. Players were encouraged to research, take real-world action, and share their experiences through blog posts, photos, and videos. The game attracted over 19,000 registered players from 150 countries, who collectively submitted more than 23,500 blog posts, 4,700 photos, and 1,500 videos.

The game's emphasis on real-world application and community engagement cultivated a sense of collective efficacy among players. Participants like Amina didn't just learn about global challenges—they became part of a network of changemakers, each contributing to a larger movement of social innovation. Critically, participants were not just handed a checklist of things to do. The majority of the actions taken were chosen by participants, who researched and decided on actions that were impactful and meaningful within their own context.

Urgent Evoke stands as a pioneering example of how interactive storytelling and community-driven missions can inspire tangible action. More recently, digital games have begun to carry this community-oriented focus forward.

SUSTAINABILITY, A HUMAN CONCEPT

"What I want to do with games is encourage local communities, because I think local communities are the driving power of sustainability" said Yaldi Games's founder Elena Höge[2] as a part of the Climate Resilience Center at the Atlantic Council's research into how games can drive change. Her studio's game, titled *Out and About*,[3] sees players identifying local plants and fungi, cooking recipes, and bringing together a community in a coastal town. Foundationally, however, this game is also created to encourage players to bring their in-game behaviors over to their physical lives, ideally joining and strengthening their local offline communities as well. After building healthy communities and community gardens in-game, *Out and About* wants to catalyze players' journeys outside the boundaries of play toward engaged community activists. As Höge notes, local communities, organizations, and governments often move more quickly than larger institutions and can drive meaningful impact at local scales, which makes them ideal targets to send players toward.

We'd love to see more game creators feel ready and able to design across the boundaries of traditional gameplay and aim for this sort of impact. We admit that success here is never trivial, however; it takes experience, or training, and perhaps tools and resources. Success here definitely requires an in-depth understanding of a theory of change, and we believe it really requires an understanding of how your game might fit into broader sustainability movements. After all, we have a lot to learn from these older movements.

The modern concept of sustainability first emerged in the late 1970s,[4] with a formal definition arriving in 1987 via the Brundtland Commission, who wrote that sustainable development is "development that meets the needs of the present without compromising the ability of future generations to meet their own needs."[5] While the game industry predates this concrete language, notions of sustainability are arguably nearly as old as the industry itself,[6] especially if we consider crossover discourse from technical, industrial, and design fields.[7] From nearly the beginning, members of these industries have called for us to do better.

Just as the call to do better is not new, the call that better is *possible*, and that our very expertise might be leveraged toward the creation of a better world[8] is also not new. These calls are good and valuable. However, no work can be done if it cannot first be imagined. This is why the next step is the work of figuring out *how* we can make this better world, in a way that applies to *now*.

Practice, Theory, and Leveraging Both for Change

As you read the chapters within this book, we hope you are guided in your reading by the theory of change that we, the editors, hold critical: that the goal of games should be to empower action, not to shift behavior.

When reading the above, you may find yourself thinking that the two words, "action" and "behavior," are generally interchangeable, and that the drawing of a line between the two is purely academic. You may simply disagree, asserting that behavioral shifts are the primary goal of any game for impact. We ask, however, as you read this book, to keep in mind a critical differentiation between the two.

Behavior is a single moment in time, often simply a reaction to outside stimulus. In the same way that a trainer can teach a dog to sit when offered a treat, a game can train a player to click on the "donate" button during a loading screen by offering the appropriate reward. However, behavior is not a sustainable goal, as their focus is on a response to stimulus that cannot be reasonably predicted to exist in the player's day-to-day. Instead, we must strive to empower players to discover and act on behaviors that are meaningful to and possible for them.

Behavior is often the target of pro-environmental interventions for a number of reasons. First, immediate, short-term behavioral shift is relatively easy to influence compared to long-term behavioral patterns. Short-term behaviors are also easily observable as a researcher can simply check the box that a person engaged in the relevant behavior. However, these short-term behaviors, especially those resulting from a response to a stimulus we know the player won't encounter in their day-to-day, do not necessarily translate into long-term behavioral shifts; otherwise known as actions.

For this reason, it can be more useful for you to consider a theory of change that prioritizes action. Action has three key elements:

1. A desire to change (either in ourselves or the world around us),
2. A belief that we are powerful enough to create that change, and
3. An understanding of how to create that change.

With these three elements, a player is not simply trained to engage in a behavior. Rather, they are empowered to take action in a way that reflects their own attitude, their own understanding, and their own perception of their ability.

In the section mentioned below, we provide a holistic overview of this theory of action, supported by evidence across the literature of environmental education, behavioral psychology, and serious game design.

DESIRE, ATTITUDE, AND ACTION

The desire or intent to act is a strong predictor of action.[9] Frequently, we talk about this with the umbrella term *pro-environmental attitude*.[10] Attitudes develop over time through exposure to nature, internal social influences, and external social norms.[11] Games can help strengthen these attitudes by fostering empathy[12] and encouraging identification with pro-environmental groups.[13] Moreover, games can mitigate negative social influences by normalizing sustainable behaviors within game mechanics.[14]

Both research and practice have shown us the power of games to shift attitudes.

Though limited in its distribution, a 2018 geo-locational game developed by a German research team asked players to spend time in their local forests, using mobile devices to engage with AR/XR elements of the forest and participate in a simulated forest management experience. The evidence clearly showed the players felt a connection to the natural spaces in their area, a key element encouraging conservation behaviors.[15]

Studies show that narrative-driven games encourage players to empathize with environmental issues.[16] The recently published and critically acclaimed *Alba* did just this, drawing players into a deep connection with a virtual space and fostering a sense of connection to that space that was very likely to translate into increased connection to real natural spaces.

Perhaps, most pointed is the ability of the culture surrounding a game to reinforce (or discourage) belief and behavior norms.[17] The power of the player community to do both was made abundantly clear in a recent study detailing the experiences of *Subway Surfers* players, whose social media community reacted starkly to the introduction of messaging to convince players to reduce meat intake.[18]

Out and About's subtle experience provides a peaceful and approachable method by which players, even those who have never left the dense urbanization of a major city, can find a connection with nature and build their desire to both connect with and protect natural spaces in their area.

THE ROLE OF KNOWLEDGE IN ENVIRONMENTAL ACTION

Environmental education research identifies four types of knowledge essential for empowering action: awareness, systematic, action-related, and effectiveness knowledge.[19] In short, if you want someone to effectively act in a pro-environmental way, they must be aware of the problem,[20] understand the systems that influence the problem and its context,[21] know what actions they can take,[22] and how effective those actions will be.[23] Effective environmental education integrates all four types of knowledge to create meaningful change.

Games are highly effective at teaching environmental knowledge by simulating real-world systems and encouraging experiential learning.[24] For example, management simulators allow players to experiment with sustainability decisions in a controlled environment.[25] Environmental education games often incorporate elements like resource management, competing priorities, and decision-making processes to reinforce systematic knowledge.[26] Moreover, collaborative and competitive game elements have been shown to enhance systems thinking.[27]

In our above example from Yaldi, *Out and About* provides players with functional action-related knowledge. The player, through their in-game activities, begins to build an understanding of flora and fauna, regional biodiversity, and ecosystem services simply by playing the game.

BELIEF AND ENVIRONMENTAL ACTION

Research has provided significant evidence that someone must believe their actions are impactful for them to follow through with any long-term behavior.[28] This holds true of groups as well, with a person being more likely to take action, if they feel they are part of a larger community taking action.[29] This is known as *perceived self-efficacy*. Games enhance self-efficacy by allowing players to simulate problem-solving scenarios, reinforcing the idea that their actions have consequences. Environmental games, in particular, have been designed to enhance perceived efficacy through real-world action integration[30] and local community engagement.[31]

Games do this in myriad ways as they are, at their core, programs built to teach a player how to do better next time than they did last time. In the simplest of examples, we can examine the core game loops of classic platformers. The player must learn to jump, then jump further, then jump across obstacles, and so on. With each progression, they gain increased confidence in their ability to complete tasks.

As a tool for encouraging sustainable actions, games support self-efficacy by allowing players to safely take actions that, in their real lives, would be considered risky.[32] Role-playing mechanics can provide a sense of agency and control, increasing motivation to engage in real-world environmental behaviors.[33] In games such as *Final Fantasy VII*, the player can experiment by roleplaying as a character involved extreme sustainability behavior and gain confidence in their own ability to take on the role of activist.

Games can also inspire confidence and self-efficacy by encouraging players to take real-world actions. In 2018, Salvatore Di Dio and his team created *Traffic O2*, a game that aimed to reduce car traffic and increase foot traffic in a major city.[34] Using simple rewards and game-based interactions, the game inspired commuters to rely on bicycles and walking, to visit areas of their city much more accessible without a motor vehicle, and to spend money at these places. The result was a user group that felt empowered, that they could bolster their community and reduce carbon without the need for major governmental involvement.

Few games more perfectly illustrate the ability of the game industry to facilitate the transition between in-game self-efficacy and real-world self-efficacy than our example *Out and About*. The game very intentionally diverts players into community efforts that can translate in-game skills into real-world actions.

THE THEORY AS A WHOLE

We recognize that, as designers and developers, researchers and curious students, artists and engineers, the above theoretical perspective can feel intimidating. When confronted with this feeling, we encourage you to remember that, in the end, it all boils down to a fairly simple approach, one that can fit into any role in the game industry.

1. Create or tap into a desire of players or professionals to have meaningful and positive climate impact,
2. Give people the knowledge and tools they need to understand and act on an issue,
3. Show them how they can succeed and encourage a belief in their own empowerment, and
4. Connect them with a community that can sustain and grow their impact-minded actions over time.

Keep these four steps in mind, and you'll rarely stray from an impactful path.

Moving Forward, Together

In many ways, the long road is more real than the destination. Our societies have structured themselves to resist change toward sustainability, and so much of our world has been shaped by consumption and consumerism that it can be difficult to even imagine a truly sustainable world. But the worthwhile pursuit of meeting the needs of the present without borrowing from the resources of the future must inevitably grapple with the ever-changing complex circumstances of both the present and the future.

Faced with a problem without a way to prove a single answer, our perspectives and approaches must shift. Rittle and Webber's "wicked" label for such problems remains useful as they remind us to refocus on the direction of work rather than the end point.[35] When circumstances continuously shift and measurement becomes chaotic, the value instead lies in re-solving continuously, aiming to move toward "better" one step at a time.

This is the work of a lifetime. Moreover, it is the work of generations.

Knowing this is freeing. If you take one thing away from this book, know that nothing in this book will solve the state of sustainability in the game industry and the world—and know that this is ok, because instead, this is a book suggesting paths toward better, and better is worth working toward. You alone cannot solve sustainability in games, or game makers, or the world, which means you are not responsible for this lofty aim. Instead, we hope this book leaves you ready to participate in the collective effort to increase our sustainability. We hope you help us reach the day when this book is obsolete because games and their makers have moved beyond us to new challenges. On that day, the work will still not be done, but the work completed will be worth celebrating and the path forward will look different. Maybe you'll have been a part of bringing us to that day. Maybe you can write the book that helps us all figure out what to do next.

What previous works in this space were expressing and refining, and what this book builds upon is a collective theory of change. A theory of change is a concrete expression of *how* and *why* actions (often called "interventions") will lead to outcomes (or "results"). The more explicitly and coherently a theory of change can tell the story of how our current situation might become a better situation, the more useful a tool it becomes to inform our priorities, decisions, and actions. The more robustly a theory of change's causal assumptions hold up in practice (i.e., does *x* action actually lead to *y* result as expected?), the more likely interventions will successfully drive toward intended results.[36]

We have one for this book, as we've described above, even as we acknowledge that our theory as editors is not a perfect monolith that captures all of the thinking in this book. Theories of change grow more robust over time through critical scrutiny, fractal adaptation to specific contexts, and constant iteration. As causal assumptions are tested, they can be refined or replaced; as procedural gaps are found, new plausible steps can be articulated. We are still figuring this out.

Broadly, theories of change are useful because they force us to check that we know where we are starting from, that we are headed in the right direction, and that we have a plan. They are the map of that long road of work toward the better world we want, and ideally, they can inform you when you are taking your next step on this road.

We hope you find the theories of change in this book engaging and useful. We also hope that you read them critically, adapt them to your purposes, and find new methods of navigating in our shared direction of sustainability.

NOTES

1 World Bank, Jane McGonigal. Urgent Evoke. 2010
2 York, Trevin, Catherine-Ann McNamara-Peach, and Ariadne Myrivili. "Gaming for Climate Action." One Billion Resilient. The Climate Resilience Center at the Atlantic Council, April 2, 2024. https://onebillionresilient.org/know-your-practical-plan-for-impact/.
3 Yaldi Games. "Out and About." Yaldigames.com, 2025. https://www.yaldigames.com/outandabout.
4 Jeremy L. Caradonna, *Sustainability: A History, Revised and Updated Edition* (New York: Oxford University Press, 2022).
5 Brian R Keeble, "The Brundtland Report: 'Our Common Future,'" *Medicine and War* 4, no. 1 (1987): 17–25.

6 York, Trevin, Catherine-Ann McNamara-Peach, and Ariadne Myrivili. "Gaming for Climate Action." One Billion Resilient. The Climate Resilience Center at the Atlantic Council, April 2, 2024. https://onebillionresilient.org/know-your-practical-plan-for-impact/.

7 Victor Papanek, *The Green Imperative: Ecology and Ethics in Design and Architecture* (London: Thames & Hudson, 2022).

8 Jane Mcgonigal, *Reality Is Broken: Why Games Make Us Better and How They Can Change the World* (London: Vintage, 2011).

9 Katarzyna Byrka, Terry Hartig, and Florian G. Kaiser, "Environmental Attitude as a Mediator of the Relationship between Psychological Restoration in Nature and Self-Reported Ecological Behavior," Psychological Reports 107, no. 3 (2010): 847–859, https://journals.sagepub.com/doi/10.2466/07.PR0.107.6.847-859.

10 Ko-chiu Wu and Po-yuan Huang, "Treatment of an Anonymous Recipient: Management Simulation Game," *Journal of Educational Computing* 52, no. 4 (2015): 568–600, https://doi.org/10.1177/0735633115585928.

11 Taciano L. Milfont and John Duckitt, "The Environmental Attitudes Inventory: A Valid and Reliable Measure to Assess the Structure of Environmental Attitudes," *Journal of Environmental Psychology* 30, no. 1 (2010): 80–94, https://doi.org/10.1016/j.jenvp. 2009.09.001.

12 Leila Scannell and Robert Gifford, "Personally Relevant Climate Change: The Role of Place Attachment and Local Versus Global Message Framing in Engagement," *Environment and Behavior* 45, no. 1 (2013): 60–85, https://doi.org/10.1177/0013916511421196.

13 Lorraine Whitmarsh and Saffron O'Neill, "Green Identity, Green Living? The Role of pro-Environmental Self-Identity in Determining Consistency across Diverse pro-Environmental Behaviours," *Journal of Environmental Psychology* 30, no. 3 (2010): 305–314, https://doi.org/10.1016/j.jenvp.2010.01.003.

14 Katherine Farrow, Gilles Grolleau, and Lisette Ibanez, "Social Norms and Pro-Environmental Behavior: A Review of the Evidence," *Ecological Economics* 140 (2017): 1–13, https://doi.org/10.1016/j.ecolecon.2017.04.017.

15 Joachim Schneider and Steffen Schaal, "Location-Based Smartphone Games in the Context of Environmental Education and Education for Sustainable Development: Fostering Connectedness to Nature with Geogames," *Environmental Education Research* 24, no. 11 (2018): 1597–1610, https://doi.org/10.1080/13504622.2017.1383360.

16 Daniel Fernandez-Galeote et al., "Gamification for Climate Change Engagement: Review of Corpus and Future Agenda," *Environmental Research Letters* 16 (2021).

17 James Paul Gee and Elisabeth Hayes, "Nurturing Affinity Spaces and Game-Based Learning," in *Games, Learning, and Society: Learning and Meaning in the Digital Age*, ed. Constance Steinkuehler, Kurt Squire, and Sasha A. Barab, 1st ed. (New York: Cambridge University Press, 2012), 129–153.

18 Clayton Whittle, Amy Rodger, Trevin York. "Game Design for Environmental Action: Lesson from studios in the Green Game Jam 2024." (Playing for the Planet, 2025).

19 Jacqueline Frick, Florian G. Kaiser, and Mark Wilson, "Environmental Knowledge and Conservation Behavior: Exploring Prevalence and Structure in a Representative Sample," *Personality and Individual Differences* 37, no. 8 (2004): 1597–1613, https://doi.org/10.1016/j.paid.2004.02.015.

20 Jing Shi, Vivianne H.M. Visschers, and Michael Siegrist, "Public Perception of Climate Change: The Importance of Knowledge and Cultural Worldviews," *Risk Analysis* 35, no. 12 (2015): 2183–2201, https://doi.org/10.1111/risa.12406.

21 Joachim Schaan and Erwin Holzer, "Studies of Individual Environmental Concern: Role of Knowledge, Gender, and Background Variables," *Environment and Behavior* 22, no. 6 (1990): 767–786.

22 Julie Ernst and Stefan Theimer, "Evaluating the Effects of Environmental Education Programming on Connectedness to Nature," *Environmental Education Research* 17, no. 5 (2011): 577–598, https://doi.org/10.1080/13504622.2011.565119.

23 Mary Pothitou, Richard F. Hanna, and Konstantinos J. Chalvatzis, "Environmental Knowledge, pro-Environmental Behaviour and Energy Savings in Households: An Empirical Study," *Applied Energy* 184 (2016): 1217–1229, https://doi.org/10.1016/j.apenergy.2016.06.017.

24 Jule Schulze et al., "Design, Implementation and Test of a Serious Online Game for Exploring Complex Relationships of Sustainable Land Management and Human Well-Being," *Environmental Modelling and Software* 65 (2015): 58–66, https://doi.org/10.1016/j.envsoft.2014.11.029.

25 Shamila Janakiraman, Sunnie Lee Watson, and William R. Watson, "Using Game-Based Learning to Facilitate Attitude Change for Environmental Sustainability," *Journal of Education for Sustainable Development* 12, no. 2 (2018): 176–185, https://doi.org/10.1177/0973408218783286.

26 Jason S. Wu and Joey J. Lee, "Climate Change Games as Tools for Education and Engagement," *Nature Climate Change* 5, no. 5 (2015): 413–418, https://doi.org/10.1038/nclimate2566.

27 Cyril Brom et al., "You like It, You Learn It: Affectivity and Learning in Competitive Social Role Play Gaming," *International Journal of Computer-Supported Collaborative Learning* 11, no. 3 (2016): 313–348, https://doi.org/10.1007/s11412-016-9237-3.

28 Albert Bandura, "Social Cognitive Theory in Context," *Applied Psychology: An International Review* 51, no. 2 (2002): 269–290.

29 Philipp Jugert et al., "Collective Efficacy Increases Pro-environmental Intentions Through Increasing Self-efficacy," *Journal of Environmental Psychology* 48 (2016): 12–23.

30 Michael S Horn et al., "Invasion of the Energy Monsters: A Spooky Game About Saving Energy," in Games + Learning + Society Conference, 2016.

31 Stephen Flood et al., "Adaptive and Interactive Climate Futures: Systematic Review of 'serious Games' for Engagement and Decision-Making," *Environmental Research Letters* 13, no. 6 (2018), https://iopscience.iop.org/article/10.1088/1748-9326/aac1c6.

32 Douglas Clark and Mario Martinez-Garza, "Prediction and Explanation as Design Mechanics," in *Games, Learning, and Society: Learning and Meaning in the Digital Age*, eds. Constance Steinkuehler, Kurt Squire, and Sasha A Barab, 1st ed. (New York: Cambridge University Press, 2012), 279–305.

33 Jan L. Plass, Bruce D. Homer, and Charles K. Kinzer, "Foundations of Game-Based Learning," *Educational Psychologist* 50, no. 4 (2015): 258–283, https://doi.org/10.1080/00461520.2015.1122533.

34 Salvatore Di Dio et al., "Involving People in the Building up of Smart and Sustainable Cities: How to Influence Commuters' Behaviors through a Mobile App Game," *Sustainable Cities and Society* 42, no. May (2018): 325–336, https://doi.org/10.1016/j.scs.2018.07.021.

35 Horst W. J. Rittel and Melvin M. Webber, "Dilemmas in a General Theory of Planning," *Policy Sciences* 4, no. 2 (June 1973): 155–169, https://doi.org/10.1007/BF01405730.

36 John Mayne, "Theory of Change Analysis: Building Robust Theories of Change," *Canadian Journal of Program Evaluation* 32, no. 2 (September 2017): 155–173, https://doi.org/10.3138/cjpe.31122.

SECTION 1

Systemic Approaches

Section 1
Systemic Approaches

4

Patrick Prax

This section will show you what systemic and collective change can mean in practice, how to connect games to collective action in practice and during design, why this broad collective approach really is necessary, and how to get started, no matter where you are.

The first section of this book takes the systemic approach to sustainability from the introduction chapters toward practical action and real implementation. This section leaves you ready to dive deeper into both the context of making games and the content of games to see how we can work for sustainability there.

First, a chapter from Patrick Prax offers a more pragmatic take on what meaningful sustainability and systemic change are in the context of games. For this purpose, it discusses how you can work in your game studio toward collective and material change, including examples and an outline of goals for this work. It then shows how that work can build the basis for even broader collaboration. This chapter also offers a number of strategies for countering greenwashing so that we can stay on target for systemic change. The strategies are presented with short stories from the author's lived experience working with this topic, both so that they are connected to reality and easier for you to share with others.

The next chapter from Aric McBay and eileen mary holowka continues that focus on collective change and action by looking at social movement games. Their chapter discusses how games can support and be made in the context of social movements. This chapter is especially interesting because it comes from and is informed by real work and participation in social movements and in experience as a games publisher. That is why this chapter provides a very real and bottom-up view, structured around examples and immediately practically useful advice. This chapter concludes with a toolkit for making social movement games that are themselves a kind of action and includes participatory design to keep the game-making process collective.

Both these chapters consider the importance of collective and systemic action, one from a more theoretical perspective and one informed by the practice of social movements. Chapter 3 by Thorsten Busch, Florence Chee, and Tanja Sihvonen then

DOI: 10.1201/9781003512400-5

gives a very comprehensive perspective on why we need this kind of massive change. Their chapter explains the economic, social, and environmental dimensions of what sustainability in and through games needs to mean. The analysis of the game industry from a business perspective offers a critical reading anchored in up-to-date political economy analysis of games as a cultural industry and can support strong arguments for a business case of systemic change. It is based on interdisciplinary research and the state of the art of critical social science in the area. If you are looking for an honest view on the sustainability issues of games that is informed by ongoing research, then this is for you. This chapter ends with recommendations for how to start addressing some of these issues on a micro-, meso, and macro-level that provide the connective tissue to the other chapters in this section.

This section concludes with Arnaud Fayolle offering pragmatic guidelines for making every game green, no matter where you are in a studio. This chapter is an absolute key to this book, because it takes the complex and tricky questions that have been discussed so far and then shows you how to start a conversation about bringing something from them into exactly the game you are making with precisely your team and specifically your players. Arnaud makes the tricky balancing act of including work for systemic change look easy with pragmatic steps to find your game's "Unique Impact Potential" in collaboration with your team. Arnaud is coming here from experience in the industry and the knowledge of how to champion sustainability in a team so that everybody can get on board.

Systemic and Collective Change in Practice— Examples and Inspiration

5

Patrick Prax

ABSTRACT

This chapter offers examples for what material, collective, and systemic change can look like in different areas of the game industry and for game makers no matter what they are working with. It makes it easier to understand, apply, and communicate the framework for meaningful climate action. This chapter offers a pragmatic step toward formulating a plan for systemic change that accounts for the possibility of greenwashing. This chapter also features a number of strategies that can help in recognizing greenwashing and that come with stories that make it easier to explain them to others.

KEY TAKEAWAYS

- Examples of what systemic change could look like for many stakeholders in the game industry and for most positions in a studio
- Relevant long-term goals for your political work and activism to keep it on track and to think clearly about where you want to be going
- Pragmatic and shareable ways to think about systemic sustainability and spot greenwashing (to a degree; good propaganda is good, etc.)
- Inspiration to forge new pathways to systemic change.

DOI: 10.1201/9781003512400-6

MEANINGFUL SUSTAINABILITY

The introduction chapter, part 1 of this book, argued that the meaningful engagement with climate change needs to:

1. move from individual to collective responsibility and action,
2. go beyond the discursive toward **material change,**
3. **resist greenwashing** that (purposefully) impedes transformative change, and
4. lead to **systemic change** in the way we produce society.

However, it is not immediately clear what systemic, collective, or material change looks like in game making, or how aiming for it should inform our actions and planning. On an even more cardinal level, it is simply difficult to think in terms of collective, material, and systemic change. When attempting to solve a problem, we do not typically suggest changing the entire system that creates the problem. Instead, we usually look for more gradual adjustments. This is especially true for sustainability questions because, frankly, it is beyond any single person's capability to imagine the radically changed society that could address the climate crisis. This reluctance to consider system-wide change is also visible in discussions with others. In conversations about issues like climate change, whenever we pitch systemic change, we are met with a multitude of critical questions. There seems to be the expectation that the alternative system we try to imagine already must be perfectly mapped out in detail and with every imaginable problem solved, or else it is not a believable pathway to change but instead a naïve pipedream. Game makers might be the best-equipped people on the planet to do this kind of world-building, but even for us, this is impossible alone. While a bias toward the status quo is not justified when the existing political and economic systems are destroying the planet, some of this skepticism is understandable. To address this issue of thinking and communicating systemic change, this chapter will make it easier to approach the topic.

The first part of this chapter gives examples of different sustainability interventions for actors in the game industry (e.g., individual developers, game studios, and bigger, infrastructural organizations) in order to show what material and collective change can mean in practice. This is not a complete list but instead represents a collection of inspiring and innovative examples of interventions that consider systemic change as a goal. They are meant as jumping-off points to help you imagine, and then share with your peers, what meaningful sustainability work can look like for game industry professionals.

The second part of this chapter will then offer a number of strategies that are meant to help you disarm attempts at greenwashing. These strategies are explained through concrete stories that help you to recognize and counter greenwashing strategies like incrementalism, the constant request for perfect answers before even the first practical change, the diffusion of responsibility, and the use of the good games can do as an excuse to continue destructive practices.

Equipped with these examples and strategies, you will be able to think pragmatically in terms of systemic change, to stay the course even in the face of greenwashing, and hopefully, you will also be able to share these complex perspectives with others in an approachable way.

Sources and Data

Because this chapter attempts some kind of overview and categorization of sustainability goals and actions, it is important to briefly discuss where the data and input for this are coming from. One of the most important sources is the collection of strategies that has been produced by the International Game Developers' Association's Climate Special Interest Group[1] and can still be accessed under www.greengamedesign.com. Other sources are policy documents produced by Project Drawdown and Playing4thePlanet. I also draw on scholarly discussions at research conferences and sustainability game jams as well as the work with my own master students who were asked to offer their views after working with the topic. The use of these sources is not an uncritical endorsement; each has shortcomings. But each holds significant and unique value for this conversation, like strategic roles in organizing support and information for other actors or allowing best practices to scale, and the intent of this chapter is to provide functional guidance, not offer critique on any of these actors or initiatives.

THE PATH TO COLLECTIVE AND MATERIAL CHANGE

If the aim of climate action is collective, material, and systemic change, then what would be a good example of that? And what do these categories really mean?

In the above diagram (Figure 5.1), I have illustrated the four different categories of change on the axis of individual to collective change and discursive to material change.

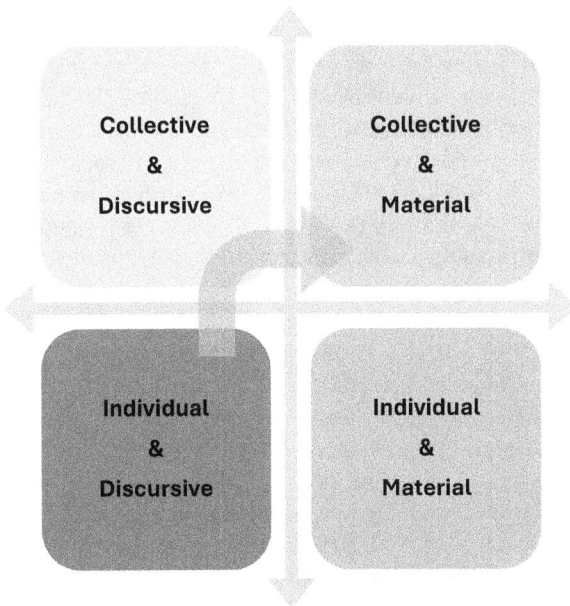

FIGURE 5.1   Categories of climate action.

These resulting four categories describe:

Individual and Discursive: Thinking about it alone.
Individual and Material: Changing what you do alone.
Collective and Discursive: Discussing with others what we could do together and organizing.
Collective and Material: Changing what we all do together.

ONE EXAMPLE IN DETAIL—FROM THE PERSPECTIVE OF A GAME DEVELOPER

Based on the above diagram, we can consider which actions different stakeholders could take in different corners of the diagram to finally reach the collective and material change. As an example of how this diagram can be used, we will consider a perspective that most of us can likely understand to some degree, that of a single developer working in a commercial game studio. If you're a single developer in a game studio, then your starting point will typically be individual and discursive (see Figure 5.2). Here, this means that you, by yourself, change the way you think about sustainability in your work and your responsibility. You could start using a carbon calculator to track and analyze your emissions. From here, the next step is frequently individual and material. You could reduce your own emissions based on the analysis from the calculator and make changes to the way you travel, commute, or use energy. Turn off the screens when going home. You could even code with more focus on energy efficiency.

However, there is no direct and obvious path from you changing your own behavior to doing this together with other people in a group without changing their minds, educating them, and getting them on board first. This means that the next step from individual material change would typically have to be toward collective and discursive change: discussing with others what we can do together. This is where you can educate other developers in your studio, build a network or group of people who care about these issues, and organize. The Climate SIG of the IGDA is such a community where you can learn from each other and think in new ways, together. Here, you might calculate the emissions for the entire studio and its games to show your climate impact and to be able to argue for which changes to make first.

The important next step to collective and material change then happens when the community (in this example, this means all the people in your game studio) change the way they are working and making games in order to reduce their climate impact materially. This could mean many things. They might work together on the code and infrastructure of the game to reduce its energy consumption. This could happen based on the emission calculations that you and your colleagues already prepared in the previous step. It goes beyond changing how all your colleagues commute and order vegetarian meals as a baseline. Considering the emissions of the game being played and reducing them is central here. Frugal coding, reduced resolution and refresh rate in menus and background graphics, default eco mode, reduced patch size and frequency

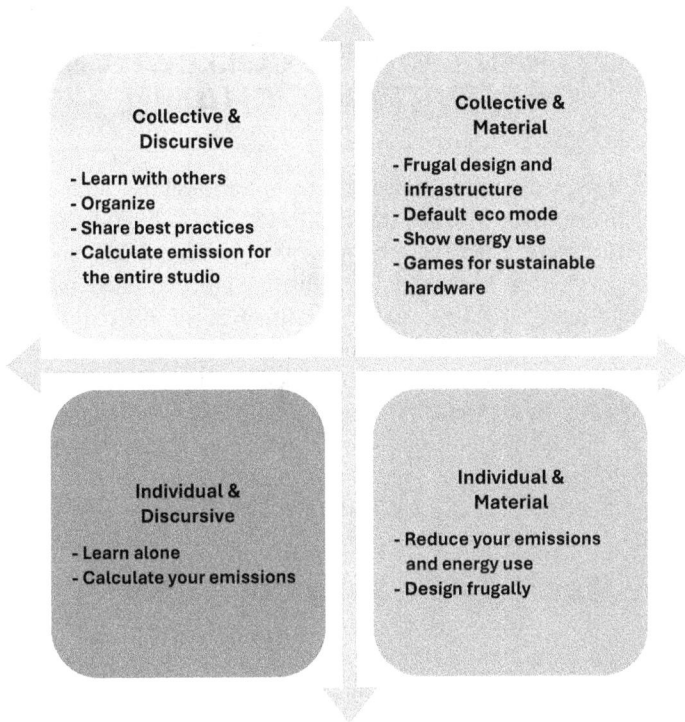

**Collective &
Discursive**

- Learn with others
- Organize
- Share best practices
- Calculate emission for
  the entire studio

**Collective &
Material**

- Frugal design and
  infrastructure
- Default eco mode
- Show energy use
- Games for sustainable
  hardware

**Individual &
Discursive**

- Learn alone
- Calculate your emissions

**Individual &
Material**

- Reduce your emissions
  and energy use
- Design frugally

FIGURE 5.2   Categories of climate action for a single game developer in a studio.

are all just the first suggestions of what could be done on the level of the studio. Eco mode specifically is an intervention that has gathered support recently. Eco mode is an opt-in or ideally default setting with the possibility to opt-out, reduction of energy use, and graphic quality by players where a game runs on purposefully reduced resources in a way that does impact the experience as little as possible. A default eco mode has, for example, been employed in *Call of Duty: Modern Warfare II*[2] where it reduces, by default, the graphic settings of in-game menus.[3]

More detailed examples of frugal coding are things like graphical sufficiency, support for low-end hardware, reduced internet traffic, optional HD assets, reduced electricity use, cleaning up data storage and cloud services, and lowering resource use when pausing the game. There is a lot of complexity hidden here and a lot of work to still be done in creating practical pathways for others to follow so that this change can scale.

The category of individual and material change is tricky. As discussed in the earlier chapter that outlined the underlying philosophy of this book, stressing individual change as the only valid way to address climate change is an effective greenwashing strategy. This does not make it so that individual change and responsibility have no place in climate action. However, it does mean that, from our perspective, the safer and more productive path to material and collective change is organizing and working together. This is why the path we champion leads over collective and discursive change.

TOWARD LARGER COLLECTIVES AND SYSTEMIC CHANGE

Once you have organized within your game studio (or if you are already leading as a studio director), the next step involves starting this process of building collective action over, now on a larger scale. Where before you were organizing developers, now you'll be organizing between studios. The end point of the first process, where one studio works collectively toward material change, is the starting point of this next process. As our scale of focus changes, our structure of collective action remains the same. This structure makes it possible to connect the grassroots actions of one individual human being to the goal of industry-wide change. We merely build and act through larger scales of collectives. To do this, we need to complete the cycle of collective and material change repeatedly, moving between scales. Yes, it is a journey, but it is conceivably possible, and this framework offers guidance for where to focus.

The Game Studio as the Individual Actor

The next step, again, requires organizing with others; at this scale, collaboration among game studios and other industry actors can achieve industry-level impact. For a game studio, the individual categories are basically the same as the collective ones for single developers. There are considerable possibilities on the level of making games to empower and educate players, and possibly even directing them to collective change.

- Individual and Discursive
  - Educate the studio
  - Calculate emissions

- Individual and Material
  - Frugal design
  - Games for older tech
  - Default eco mode
  - Make games to empower and educate players
  - Showing energy cost to players
  - Collective achievements to save energy with eco mode.

The collective actions here are central for taking that next step and changing the material reality of making games. When many game studios come together, they can formulate best practices and influence the way things are done in the industry and change the practice and reality of game making. Where one studio faces pressure to conform to an industry norm, many together can change the tide.

- Collective and Discursive
  - Implement standardized carbon emission and energy use calculations and reporting

- Educate the industry collectively
- Share best practices and create white papers
- Create trade unions to share this information and political work
- Hold trade unions and lobby organizations accountable for sustainability goals
- Present a united front to hardware manufacturers and distribution platforms.

- Collective and Material
  - Use the emission data and reduce emissions and energy use
  - Put pressure on platforms and policymakers to change the infrastructure of game production

Where this second level here considers game studios as actors, and the third looks at infrastructural organizations.

Infrastructural Organizations

This process of creating a bigger collective can be repeated one more time, incorporating collaborations with industry lobby organizations like the "Entertainment Software Association" (ESA) or for example the Swedish or Finnish game industry representatives. Such a collective would have the ability to negotiate with some of the most powerful stakeholders, such as big publishers, digital distribution platforms, and console manufacturers (like Steam, Google, Microsoft, and Sony). Game industry lobby organizations already influence legislation and policy, but if they were integrated into a collective movement for material change and held accountable to this goal, they could support and inform policy on sustainability issues that address the material and systemic conditions of game production. The most obvious examples of outcomes would be building meaningful consequences to the Corporate Sustainability Reporting Directive (CSRD) based on standardized sustainability reporting. Collectively, we could establish sustainability requirements for public game funding, something currently under discussion in the European Union where there is a real possibility for long-term systemic impact. We could even advocate for setting legal energy use limitations for games, creating an internationally equal playing field for competition where studios that choose to build less energy-intensive games are not at a disadvantage.

Beyond industry lobby, the category of *infrastructural organizations* also contains the already named digital distribution platforms like Steam or Google Play, gaming technology producers like Sony and Microsoft, and even policy makers like the European Union. These organizations have the power to reshape the way we make and play games. They can change the scaffolding of the cultural industries that games are a part of. Their actions could directly lead to the systemic change that is our goal. Because these stakeholders described here are so different, it is difficult to prescribe aims that fit all of them. Separating them and offering more detailed work on how to address them specifically would be useful but is beyond the scope of this chapter and something we still need to figure out. However, below are some suggestions for what material and collective goals for these stakeholders could look like.

- Individual and Discursive
  - Enforce carbon emission tracking

- Individual and Material
  - Mandatory eco mode
  - CO2 labels for games
  - Mandatory visibility of energy use

- Collective and Discursive
  - Educate entire sectors
  - Share practical guidance and research-based methods

- Collective and Material
  - Energy budget for an entertainment/gaming hour, alternatively Carbon Taxing
  - Take-back rules for e-waste
  - Ban the export of e-waste
  - Ban conflict minerals
  - Ban unnecessary streaming (of ads)
  - Regulate planned obsolescence of hardware.
  - Slow the tech arms race, for example by only allowing hardware made from reclaimed materials.

These suggestions purposefully stressed the material elements of game production, including hardware production and use. Something like an energy budget for an hour of entertainment could have clear knock-on effects on the trajectory of global computation hardware production and resources use in programming, thus conceivably impacting not just the game industry but the technology sector.

Supporting Other Collectives

This structure also makes it possible to consider where our actions can help other stakeholders on their path toward collective and material change. That is of course a core part of the sustainability work of games already: making games to inspire our players to work for sustainability. However, we also need to consider that we should inspire players to aim for systemic and collective change. An example of a game that shows material consequences of play to players is the game Ultros.[4] The game shows the estimated energy consumption of play to the player when selecting the graphic settings (see Figure 5.3), indicating for example that a high-end resolution will use drastically more energy than a medium setting.

Another interesting example of this line of thinking is to create achievements for players that play in eco mode together. If players as a group unlock collective

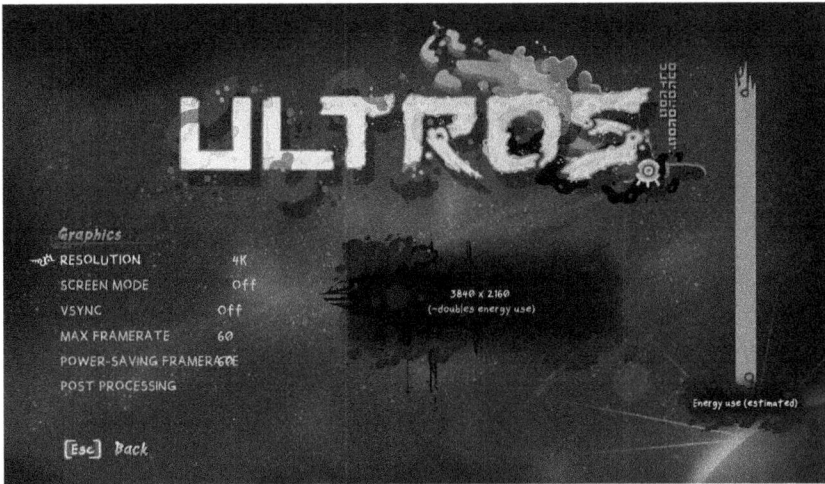

FIGURE 5.3   The graphic options menu of the game Ultros showing the power use of the game depending on the graphical settings chosen (Hadoque, 2024).

achievements by saving energy in their play, that means that the design frames their individual actions, playing in eco mode, as a part of a collective movement. This way players are pointed toward collective action. Many chapters in this book show a way to do this kind of work, for example Chapters 6, 8, and 19. To reiterate, these points are not a proven and evidence-based roadmap on what should be done in which order, but suggestions for goals around collective and material change.

The point here is to help us all together to think in terms of systemic change.

STRATEGIES TO COUNTER GREENWASHING

The third requirement for meaningful engagement with climate change as set out at the start of this chapter is a resistance to greenwashing. There is no generalizable tool to categorize and "disarm" any possible type of greenwashing, but it generally involves misleading stakeholders by overemphasizing environmental benefits and/or understating negative environmental consequences.[5] Another element of greenwashing that is discussed in Chapter 2 is the individualization of climate responsibility.[6] That is why instead of attempting to define and counter any possible kind of greenwashing, this second part of this chapter shares four non-fiction short stories based that are based on our experiences of different kinds of greenwashing. These stories address ways of greenwashing like incrementalism, the constant request of perfect answers before even the first practical change, the diffusion of responsibility, and the use of the good games do as an excuse to continue the destructive practices. They serve as an example of this kind of greenwashing and lead to strategies that help you to recognize and hopefully disarm them.

Story 1: Is This the Best They Can Do?

During the Swedish Democracy Festival 2024 I participated in a panel called "Can games save the planet?" The panelists were the chair of the Swedish Society for Nature Conservation, the sustainability office of one of the biggest Nordic game publishers, the project lead for a Swedish game industry support organization, the CEO of an indie studio, a member of the Swedish Game Industry lobby organization, and me as a professor of a game design department. We are a group that combines enough power and knowledge in the room to start creating real change right then and there. We spent an hour outlining some sustainability issues and stressing that it would be important to do something.

This story is meant to account for the kind of greenwashing in which powerful organizations engage with climate change and want to be seen doing so. However, instead of committing to truly impactful actions, they pursue the minimum effort or some kind of incrementalism in which their role is to motivate others to do the real work. Here, in this panel, it could have been possible to write a white paper, pull relevant experts together, set up aims and a process for the Swedish game industry to create legislation with industry and academia involved, and commit resources. Instead, we talked for an hour in quite general terms and then left. This is not acceptable. And, based on the Game Developer presentation of the industry's view on our sustainability efforts, we know that it is not.

> Indie studios aren't the problem. Every indie could go carbon-negative and it wouldn't change a thing. The real culprits are corporations like Amazon and Microsoft who run massively wasteful data centers.[7]

This story is going in the same direction as the short opening vignette at the start of the introduction part 1 of this book, in which a participant in a sustainability game jam explained that their sustainability game was a part of the greenwashing of Meta. That case was even more extreme. Meta is one of a handful of the most powerful organizations on the planet. They could use their considerable platform to create real change. They could lead conversations around real policy and offer expert advice and industry insights and feasibility and smart regulations. It is easy to see the incremental work of large organizations like this one and walk away with the impression that they do care and are at least doing something. However, these initiatives are not aiming for the meaningful systemic change we need. In this situation, and when confronted with the sustainability work of a powerful actor, we need to see this incrementalism and hold them accountable in a way that considers their very real power position.

Here, it is useful to ask: "Is this the best they can do?"

Story 2: Are We Bottom-Trawling?

When discussing sustainability with my peers in sustainable development who work with fishing in the Baltic Sea, these very mild-mannered researchers can get positively incandescent whenever we reach the topic of bottom-trawling. Bottom-trawling is a fishing technique that pulls the nets over the sea floor, increasing the catch somewhat but destroying the ecosystem to a point that takes years to repair itself.[8] It is so damaging that it quantifiably hurts already next year's catch. Regulating fishing rights is notoriously complicated with many stakeholders on the line, including concerns about the way of life of local fisheries, diets, and industry. In this area, it is hard even for researchers to prescribe a specific course of action. However, nobody with any knowledge in the area supports the use of bottom-trawling. Still, it is permitted and used by Sweden (and all other countries) in the Baltic Sea. How about us working with sustainability in games. Are we all also ignoring actions that we could take right now? Are we doing things that we know we should stop right away? Are we still bottom-trawling?

The point of the story is to offer a way to address a kind of critique toward sustainability efforts that use the complexity of the problem as an excuse to stall even clear first steps in the right direction. Using this concept of bottom-trawling on the game industry and game making, we can identify things that we can clearly do better right now and then push toward doing them. Even in the face of a complex problem like climate change, and even though we do not have a perfect solution ready yet, there are clearly things we can do better already now. This strategy is meant to help you identify what should be the uncontroversial first step for anybody who honestly wants to help.

This is where it helps to ask: "Are we bottom-trawling?"

Story 3: Are We Really Just a Powerless Part of the System?

When discussing with an acquaintance from a studio how game makers could work to make games more sustainably, I suggested as a long-term goal that we could make eco mode the default on some institutional level. Their response was that this should not be done. The decision about which games thrived and profited should, in the end, be taken by customers and the market, and players wanted the most advanced graphics and processing possible. I tried to follow that logic of bottom-up influence and suggested that maybe the game industry could put pressure on hardware producers to make more energy-conserving machines. This was also rejected for the reason that this sort of upward pressure would not be effective on a systemic level.

This argument against working for change in a twisted way recognizes the need for, and complexity of, systemic change and the centrality of games and tech, and it uses these to defend inaction. This perspective sees it as impossible to impact players and customers downstream, framing game makers as powerless and giving all the power to direct change to the collective of customers and players. At the same time, this perspective also denies that same kind of collective agency of game makers influencing hardware producers further upstream. This is a contradiction and appears to be a greenwashing tactic to remove agency, and with that responsibility, from game makers. Of course, it is correct that none of these actions are trivial or easy. However, in this framing, game makers, a group holding a strategic position in the center of many of the systems that need to change, are cast as powerless and just a part of a system. Instead, we could think of ourselves as having an important role in leading change.

This is where it helps to ask: "Are we really just a powerless part of the system?"

Story 4: Why Should It Here Be Relevant to Our Climate Impact That Games Also Do Good?

> A central point of the report of the Swedish Game Industry about sustainability and games is that we should not only consider the negative carbon footprint of games but also need to consider their positive Handprint. A Handprint here is an umbrella term for all kinds of positive impact that games are making, for example in contributing to the development of tech or in making tools for other industries to optimize their work and reduce emissions. This category is not the same as Voice, which stands for the positive impact of the message of games. Together, the chapters on Handprint and Voice make up 39 pages in the report and only nine pages on the actual industry footprint. Those nine pages come to no tangible conclusions on what systemic steps, policy changes, or industry-wide best practices should be considered.

This example comes from a report mostly written by an industry organization. This unfortunately means that right now it cannot be expected to suggest steps toward systemic change, even if in this case doing so was part of the funding requirement for the project. It is a useful case for two reasons: First, it shows that there is a real need for making sure that our sustainability goals are also reflected in the work of the organizations that represent us and that could be an amazing tool for collective and systemic change. Second, and more relevant here, using emission savings in other industries or areas of society as a kind of counterbalance to games is not legitimate. For this argument, let's assume that games technology powers the visualization of an industrial structure so that it can be built in a way that conserves a considerable amount of energy. That is great! However, it is unclear why these emissions reductions should be considered to belong to games or the game industry somehow, and not to other tech- or planning stakeholders or the polluting industry itself that has just gone greener. A sensible argument here is that games and the skills and know-how that are part of their production should not be discarded because it can be useful in many areas. This is correct and might need to be said. However, nobody (relevant) suggests purging the game industry and all related skills from the face of the Earth. The fact that game making cultivates transferable skills that can be used for sustainability work elsewhere does not mean that those emission reductions are even relevant in a conversation about what the game industry needs to do.

Whenever the argument is made that games and game makers also do good and that this is a reason for why games get to keep polluting, it helps to ask: "Why should it here be relevant to our climate impact that games also do good?"

LIMITATIONS

As explained in Chapter 3 in the introduction, making the game industry and our society sustainable is an ongoing project with a timeframe of decades. Here, I have provided ideas and suggestions to inspire and motivate, in the hope that others will come up with more ideas that are, among others, better contextualized and actionable in specific historical, cultural, and national environments. That means that this is not a complete list in any way, but a starting point.

The second limitation is that the topic of systemically changing society is vastly complex and contextual to the point that we cannot claim to be experts in this or fully understand the process. Again, this is a pragmatic step toward formulating a plan for systemic change that accounts for the possibility of greenwashing. Herein also lies the most important limitation: Guarding against good greenwashing is difficult, also because it can be impossible to distinguish from the outside between a systemic project in the starting phase and incremental greenwashing. Even conceptually, the argument here is not clear yet. Making a sustainability game with a systemic message and organizing with other developers could be a great collective approach to material change if seen on a studio level, or it could be seen as Meta greenwashing. The shifting context changes the outcome. As a researcher, this means that the strategies and analysis presented here are heuristic. They are (hopefully) good enough to be useful and to inform pragmatic action, but they are not a satisfactory description of the complexity of reality.

All this is to say, these strategies to think with and categorizations are not foolproof at all but need to be applied with honesty and care for the context. Using them without care, it is easy to go too far, discrediting valid sustainability work, or not far enough in calling out greenwashing. This is also the best I can do right now.

CONCLUSION

The aim of this chapter was to help us to think pragmatically in terms of systemic change and to share these complex perspectives with others in an approachable way. The first part of this chapter showed what working for collective and material change in your studio can look like and, extrapolating from that example, showed what we need to do to reach systemic change: go toward the collective and material and when we have convinced a studio, use that as the starting point to change more studios and organize those, going all the way up to infrastructural organizations like the ESA, Steam, Microsoft, and the EU.

The second part of this chapter offered four strategies to counter different kinds of greenwashing like incrementalism or the constant request of perfect answers before even the first practical change. These strategies will help us to stay focused on the collective and material change we are working toward instead of getting lost in the details to the point that we do not act at all or are being defeated by unreasonable expectations on us as change-makers.

This systematic way of thinking can be difficult and explaining it to others is even harder. This chapter explains the strategies to counter green washing in short stories from our own lived experience to get this perspective across in a relatable and human way. These or similar stories can also be used by you as an entry point for that conversation with your co-worker about collective material change in game production. Because one thing should be clear at this point: Systemic change is something that we need to do together, and doing it requires us all to involve and empower others along the way.

NOTES

1 Alenda Y. Chang, "Change for Games: On Sustainable Design Patterns for the (Digital) Future," in *Ecogames*, by Laura Op De Beke et al. (Nieuwe Prinsengracht 89 1018 VR Amsterdam Nederland: Amsterdam University Press, 2023), https://doi.org/10.5117/9789463721196_ch01.
2 Infinity Ward. *Call of Duty: Modern Warfare II.* Activision. PC, PS4, PS5, Xbox One, Xbox Series X/S. 2022.
3 danielmjacobs, "Gaming Sustainability Call of Duty Case Study - Microsoft Game Dev," accessed April 17, 2025, https://learn.microsoft.com/en-us/gaming/sustainability/case-studies/case-studies-cod.
4 Hadoque. *Ultros.* Kepler Interactive. PC, PS4/5, Xbox Series, and Nintendo Switch. 2024.
5 Sebastião Vieira de Freitas Netto et al., "Concepts and Forms of Greenwashing: A Systematic Review," *Environmental Sciences Europe* 32, no. 1 (February 11, 2020): 19, https://doi.org/10.1186/s12302-020-0300-3.
6 Michael F. Maniates, "Individualization: Plant a Tree, Buy a Bike, Save the World?," *Global Environmental Politics* 1, no. 3 (2001): 31–52; Jennifer Kent, "Individualized Responsibility and Climate Change: 'If Climate Protection Becomes Everyone's Responsibility, Does It End up Being No-One's?,'" *Cosmopolitan Civil Societies: An Interdisciplinary Journal* 1, no. 3 (n.d.): 132–149, https://doi.org/10.5130/ccs.v1i3.1081
7 Game Developers Conference and GameDeveloper.com (2023:24). 2024 State of the Game Industry.
8 Antonio Pusceddu et al., "Chronic and Intensive Bottom Trawling Impairs Deep-Sea Biodiversity and Ecosystem Functioning," *Proceedings of the National Academy of Sciences* 111, no. 24 (2014): 8861–8866.

Opportunities for Impact Games in Social Movements and Campaigns

6

Aric McBay and Dr. eileen mary holowka

ABSTRACT

This chapter defines the concept of "social movement games" by considering how digital games can provide the opportunity to learn and practice skills of social change, such as persuasive connection, protest tactics, and long-term strategy. Based on interviews with 11 game developers, activists, and researchers, this chapter first explores ways that games can contribute to social movements and then offers tools and guidelines for developing your own social movement game.

We believe that digital games can provide the opportunity to learn and practice skills of social change, from persuasive conversation, to protest tactics, to long-term strategy. Focusing on what we call "Social Movement Games," this chapter shows how games can contribute to social movements as well as offers some tools and guidelines for creating your own.

KEY TAKEAWAYS

- A definition of "social movement games" based on real-world examples
- A way to make games as a part of and supporting a social movement
- Perspectives on what social movement games can and should accomplish
- Insights from game-makers who do this already

DOI: 10.1201/9781003512400-7

INTRODUCTION

In 2010, the Cook Inlet Tribal Council (CITC), a Tribal nonprofit organization serving Alaska Native people in the Cook Inlet region, began a collaboration with E-Line Media to preserve and share the stories, culture, and traditions of the Iñupiat. This process involved a close partnership with Iñupiat elders, cultural leaders, and storytellers, who provided insight into the traditional narratives and history of their people. The project's goal was not only to create an engaging video game but also to serve as a platform for raising awareness about Iñupiat culture, bringing it to a global audience in an interactive way. In our interview with Alan Gershenfeld, co-founder of E-Line, he explained to us that the CITC "wanted to do a commercial game, so their youth could see their culture authentically represented in a popular medium that they engage with."[1]

The development team and Iñupiat community created an "inclusive development process" to ensure the game authentically reflected Iñupiat traditions, focusing on aspects such as their connection to nature, the importance of family, and the spirit of cooperation. Together, the organizations instilled this message in the gameplay itself, where the main character, a young Iñupiat girl named Nuna, must cooperate with her arctic fox companion as they journey through a harsh Alaskan landscape, uncovering the wisdom of their ancestors. The resulting game struck a chord with players and developers around the world.

Launched in 2014 as *Never Alone (Kisima Ingitchuna)*,[2] the game has since been downloaded more than 15 million times and won awards including a Peabody, a BAFTA, and Game of the Year at Games for Change.

Gershenfeld explained that although they didn't set out to build a movement through the game, "a movement *emerged* around it."

> We continue to be amazed at the number of people we've influenced, from developers—Indigenous and other underrepresented communities that are embracing video games because they were inspired by *Never Alone*—to a wide variety of individuals who were touched by the game—not only learning about Alaska Native culture, but also reflecting on their own culture.[3]

The impact of *Never Alone* echoed beyond just the game.

Gloria O'Neill, President and CEO of CITC described *Never Alone* as

> so much more than a video game. It's about sharing Alaska Native culture with the world through our traditions and stories. It's about meeting Indigenous youth where they are and ensuring that they see their endless potential reflected in positive, strengths-based ways across modern media. It's about coming together as an Alaska Native community to build an innovative model of inclusive development. *Never Alone*'s extraordinary success showed us the true potential of combining modern tools like video gaming with authentic Alaska Native stories and values.[4]

Never Alone's financial and global success was enduring, but the impact it had on the local community was just as important.

Inspired by success stories like *Never Alone*, we believe that digital games can provide opportunities for social change and movement-building.

SOCIAL MOVEMENT GAMES

Many impact games seek to address a problem by raising *awareness* about it. However, this is rarely sufficient. Author Aric McBay's nonfiction books have emphasized that we need more than *knowledge* about a problem in order to address it. If we want to make a fundamental change in society, we also need models for action and connections with other people we can talk, learn, and work with.[5] Game developers can help provide those things by making or planning games *within the context* of real-world social movements.

In this chapter, we begin with a few key premises:

- Transformative change happens because of *social movements*, not isolated individuals.
- Progressive social movements are *emergent and participatory*, not top-down (and this can be reflected both in how we make games and the game mechanics themselves).
- Social movements have specific *needs* that must be fulfilled for them to succeed.
- Games can help fulfill many of those needs.

We also explore several related questions: What are some specific needs of a social movement? How can games help boost social movements? What are models for doing so, and how can we implement those practices *well* in our studios and partnerships?

For this chapter, we interviewed 11 game developers, activists, and researchers beginning with Italian game designer Paolo Pedercini (*Molleindustria*), who is known for his generally small, experimental, and political games. His work *Casual Games for Protestors* makes the case that "participating in social change should be exhilarating, social, intellectually and physically stimulating, liberatory and fun. Games can help craft those collective experiences."[6] When we spoke to video game writer and narrative designer Meghna Jayanth (*80 Days*,[7] *Sunless Sea*),[8] she had a similar message:

> I think there are all sorts of pleasures—like joy and fulfilment and satisfaction and meaning and purpose—that people are really craving. I think so often the way we talk about climate change or activists is from this real position of moral purity, or that it's boring, or that everyone is holier than thou. I think there's something really useful about saying, 'No, there's actually *joy* to be had in this kind of work.'[9]

Jayanth invoked the abolitionist concept of "rehearsing the future," adding that "games are a really good place for people to rehearse different possible ways of interrelating with each other and organizing communities."[10]

We have seen games succeed as a method of recruitment for the military, but we believe it has the potential to recruit for and develop alternative futures.[11] But, what does it mean to "rehearse the future" through game design and to do so *collectively*? And why are movements important in the first place?

WHY WE NEED SOCIAL MOVEMENTS

Social movements are responsible for most of the positive, transformative change in human history. Our fundamental rights and freedoms were won by long-term, organized, *collective* struggles against oppression, including the women's suffragists, unions and labor movements, anti-slavery abolitionists, anti-colonial liberation movements, and many others.[12]

We owe so much to these movements. And if we are able to overcome the challenges of the climate emergency and modern authoritarianism, we will have current and future movements to thank.

Yet, social movements are underrepresented in video games and video game worlds. (When they are present, they are often *mis*-represented.) For a variety of reasons—some practical, some cultural, and some structural—the heroes of video games are mostly *individuals*, often with super-human capabilities.

This makes for a dramatic story. But lasting change in real life rarely happens because of one person acting alone. "What force on earth is weaker," asks the union anthem Solidarity Forever, "than the feeble strength of one?" Transformative change mostly happens because of the collective action of individuals acting through social movements around a common cause.[13]

Social movements can take many forms; there is no universal model. They are usually collections of different kinds of groups, constellations of overlapping organizations, from associations to revolutionary fronts, unions to social clubs, all embedded in a grassroots base of supporters, activists, and sympathizers. Even though some activists may be more publicly visible than others, there is rarely any single central leader.

Within a movement, there is often a great diversity of opinion about goals, tactics, and strategies. Individual people and groups choose whether or not to participate in a given action—say, a protest or a sit-in—based on their own calculus of ideological alignment and solidarity, of risk and reward.

Understanding how social movements work is critical, not only for making change but for designing games and choosing core mechanics that reflect reality. Decentralized movements are different from the top-down, management models provided by tactics and strategy games that *do* involve large numbers of characters (rather than focusing on individual heroes). We see this in games too: In *Age of Empires*,[14] a worker or peon goes where you click and does what you order—you don't have to convince them that the cause is just or that the reward is worth the risk.[15] And your peon is not going to lay down their tools to go on strike for better working conditions. (Exceptions exist: games like 11-Bit's *Frostpunk* do require you to manage morale and attend to worker safety).[16]

Protests and demonstrations are obvious physical manifestations of a social movement. But those are the expressions of organizing, planning, and a larger and less visible structure—much as mushrooms are merely the fruiting bodies of larger (but hidden) mycelial networks.

To make lasting change, social movements have specific needs. They must do some or all of the following:

- **Shift public opinion** to gain support and build awareness of a problem and communicate key messages.
- **Recruit** new members and supporters.
- **Build networks** and relationships, and **organize** themselves effectively.
- **Train movement members** in key skills.
- **Generate political force** and pressure decision makers.
- Increase **safety** of participants and of participation.
- Develop **shared strategic understanding** and insights.
- Offer activists and organizers **support and respite** from the challenges of change work.

But what does it look like to incorporate these principles into our game design?

WHAT CAN SOCIAL MOVEMENT GAMES DO?

When we spoke to game director, designer, and writer Jennifer Estaris, she argued that "game design tools are quite useful for thinking about how we progress in movements."[17] Further, they have the potential to offer training in collective action skills such as disruption, skill-building, and community-building.

This chapter proposes an early and exploratory definition of social movement games as games which do all or some of the following:

- **Recognize and archive** the existence of real-world movements for change.
- **Illustrate how change is made** in our world, including the importance of *disruptive tactics* that make business-as-usual more expensive and difficult.
- Allow players to **practice skills needed in movements** for change, ideally through the game's core mechanics (rather than simply having a superficial veneer). For example, in Joel Jordon's game *Boss Battle*, players practice organizing together against their boss.[18]
- Invite players to **explore and imagine ways to create change** and fix real problems (what adrienne maree brown calls "science fictional behavior").[19]
- Build a **real movement's capacity for change**, perhaps through gathering resources/donations for the movement or strengthening the movement skills of players. For example, *Palestine Skating Game* has continued to raise thousands of dollars for colleagues in Gaza and Lebanon.[20]
- **Connect players** with real movements, communities, or opportunities to build real-world relationships (and not replacing those relationships with digital simulacra).
- **Practice ethical and cooperative work** throughout the creation of the game. The very process of creating a game can itself be impactful. Unhealthy work environments are normalized in the game industry, but the social impact of a game depends upon the health of those creating it.
- **Collaborate with social movement organizations and developers**, as seen with *Never Alone*[21] and *Atuel*.[22]

One of the central tenants of social movement games is ***collectivism* and the act of building and strengthening community**. Although a few of the developers we interviewed tended to work alone, collaboration, and partnerships seemed to be key for maintaining sustainability and optimism in the development of social movement games, whether that was in a collaboration with a social movement organization, or working cooperatively in a smaller team. What Models Exist for Social Movement Games?

Several potential models exist for social movement games:

Entirely new games built from scratch such as *Oiligarchy* by Molleindustria,[23] *Sweatshop* by Littleloud,[24] *Saltsea Chronicles*,[25] and *People Power: The Game of Civil Resistance*,[26] which was made in association with the International Center on Nonviolent Conflict. Brand new games offer great potential, since they can be designed from the ground up for social impact, and can include social movement mechanics which are not present in all games.

Latin American coop studio Matajuegos has created several games in collaboration with movement organizations. Their award-winning game *Atuel*, about the Atuel River Valley in Argentina, features the voices of real people speaking about their experience and knowledge of the river.[27] And their game *Union Drive* was developed with UNI Américas and Global Labor Justice-International Labor Rights Forum.[28]

Game Mods are a way to bring social movement content into games with already extant higher production values, but with a smaller investment of time and resources. Minecraft has been an especially rich platform thanks to its support of constructive and collaborative mechanics. Examples include Greenpeace Brazil's *Minecraft Save Amazônia*[29] and the Climate Resilience Center's *Heat Wave Survival* and *Power Grid Hero* mods.[30]

Game Events and Activations such as 11 Bit Studios adding special street art or DLC to *This War of Mine* and raising money from sales for War Child and Ukraine relief, and social movement protests within games such as *Animal Crossing*,[31] *The Sims*,[32] *Grand Theft Auto V*,[33] and *Minecraft*.[34]

Social impact game jams, scholarships, or bundles like "Towards a Free Palestine,"[35] the Games for Change scholarship,[36] the yearly Queer Games Bundles, or the Just Play Jam further encourage the creation of games for social movements.

GAMES AS A TOOL FOR SOCIAL MOVEMENTS

Games can be a tool for advancing social change; they have a special niche because of their interactivity and ability to simulate complex systems. We don't expect games to make change by themselves any more than we would expect books, films, or podcasts to cause change by themselves. Change happens because of *people*, and mostly because of people organized into movements. While games alone cannot create a revolution, they can be used for advancing social change, as well as encouraging collective action.

A number of our interviewees referenced drama theorist and political activist Augusto Boal during their interviews as an inspiration, particularly for his framing of art as a political tool. Indie game developer Joel Jordon, who we interviewed, writes in her Different Games presentation:

> As I mentioned, art that makes political commitments won't be effective for everyone, and the point of making it isn't to change the minds of people who benefit from the way things are... But if they agree with the argument, and if they see that others do as well, and if they learn to work together with them, then it will be a forum for them to sharpen their thought and their action.[37]

Games offer an opportunity for political organizing and action. There is a wide array of knowledge from developers who have already experimented in making these interventions. The following toolkit aims to pool some of this collective knowledge so that those creating social movement games can learn from what already exists and dig even deeper.

TOOLKIT—MAKING SOCIAL MOVEMENT GAMES

The more we share knowledge on how to do the work of creating socially impactful games, the less developers have to constantly reinvent the wheel. More research and documentation in this area are needed, including the preservation of some of the games that risk being lost to time and more tools for tracking the impact these games can have.

Above, we've explored some of the social movement games that have been created and what we believe makes a game a social movement game. Based on our preliminary research and interviews, we propose the following as our starting toolkit for *how to create* social movement games:

When Making the Game *is* the Impact

In our interview with award-winning narrative designer Hannah Nicklin (*Saltsea Chronicles, Mutazione*), she told us that "making games is the process, it's not the product. And the process comes back to people."[38] Gershenfeld similarly spoke to us about how important the process of making games is if the goal is truly inclusive development.

Focusing on the impact of the *process* is not irrelevant. The creator of the *Solar Server,* Kara Stone also emphasized the importance of supporting unionization and worker's rights, considering shared ownership, and thinking about how you want to pay yourself/others when trying to make an impactful game.[39]

We believe it's possible for both the process *and* the product to have an impact. In our interview with Estaris, she split the impact of social movement games into three categories.[40] Those are:

- the social movements we depict in games
- the social movements our organization or team participates in through our values and practices
- the social movements we engage in as developers *outside* of game development.

As a game developer interested in social movement games, it is worth considering all three areas and how best to use your time to focus on each of these. Where can you have the most impact?

Keep Things Simple and Intentional

Although some socially impactful games, like *Never Alone*, may have large budgets from investments, grants, and collaborations with other organizations, the majority of social movement games are made by small teams of indies.[41]

Paolo Pedercini emphasized to us that impactful games don't need to have big budgets. "It's time to go back to the idea that games can be made by solo developers or very small teams," he told us. "If you want to make a game, you can still do it by yourself," because of the growing accessibility of game-creation tools. "So just start making your own game."[42]

Focus on creating the Minimum Viable Product of what you're trying to create, aka: the simplest version of your game that can be made while still achieving its goals. This can be iterated on with future time or resources, while still including all the core components.

Creating smaller, lo-fi games not only allows for quicker production cycles but, as Abraham mentioned in our interview, it is also an excellent way to make games in a more environmentally friendly way. For example, Stone (*Ritual of the Moon, UnearthU*) created the *Solar Server*, a solar-powered web server set up from her apartment balcony, to host low-carbon video games.[43]

You don't have to tell huge stories. In our interview with Nicklin, she described how she approaches things on "quietly radical" stories—the ones "that we have that we feel are small, but actually are big because we share them all [...] We're all as rich and beautiful and complicated and terrible, and full of potential as the next person."[44]

You can also **scale down your audience**. As Abraham reflected in our interview with him:

> It's still kind of possible to make incredible games that resonate with people. [...] You need to really scale down to individuals [and] understand their context [...] If you're trying to persuade someone about climate things you need to know what their starting point is. So you can build the steps towards them.[45]

Estaris told us that her approach (as Game Director of Monument Valley 3 at Ustwo) is to respect the player's time, and to make a game only as long as it needs to be. For example, the *Monument Valley* games are designed to be accessible by a wide range of audiences, including first-time gamers.[46] Being very intentional about your audience is also important for community-building and collectivity, which are key components of social movement games. Jordon brought her game *Boss Battle* specifically to organizing groups, which led to conversations after the gameplay had ended about workers' rights and capitalism.[47]

Clearly and *Collaboratively* Agree on Your Impact Goals and Values

The "games industry" at large has been predominantly focused on project development over studio development or collectivism.[48] In order to create an impactful project, however, it's necessary to think about how you're making that project, with *whom*, and for what reasons. Developing actionable values collaboratively with colleagues and allies is a critical step in the process.

In our interview with Gernshenfeld, he described the process that the *Never Alone* team went through to develop an inclusive development process before collaborating together. This included a process of decision-making (playing to each other's strengths) and conflict resolution.[49]

The Baby Ghosts Studio Development Fund (which author eileen mary holowka co-directs) walks early studios through methods for clearly and collaboratively building these impact goals and values so that teams can work in a truly cooperative way.[50,51] We've included some of these tools here to help build intention into your goal of making social movement games:

- **Name your team pain points** (such as challenges with decision-making or capacity issues) as soon as you can and as honestly as you can so you can address them.
- **Name your values first individually and then as a team** to see if you are all on the same page. Articulating these values clearly and revisiting them regularly to see if they are being upheld and need changing. Gernshenfeld also recommended discussing your shared risk tolerance.[52]
- Instead of focusing on *what* you do (making games!), **start with the *why*** you're hoping to address. What change do you wish to make? Once you've established that, build backward to *how* you're going to address that change and then, finally, *what* you actually create. Once you know your values and your *why*, it's important to **translate these into tangible goals**. You can do this by identifying your short, medium, and long-term goals in a "Theory of Change" or what Weird Ghosts calls a "Results Flow." Weird Ghosts has free resources on how to do this at: https://learn.weirdghosts.ca/impact-tools/results-flow
- Clearly articulate your short-, medium-, and long-term goals to **create indicators for tracking impact**. Oftentimes, game developers create impactful works but have no way of measuring what kind of change their games make. Gameplay analytics, audience reactions, carbon emissions, and other measurements can tangibly show you the impact of your work.[53]
  - Although the team behind *Never Alone* had a Theory of Change, Gershenfeld told us that he regrets not implementing a tool to track how many players chose to engage with the unlockable video "Cultural Insights" featured in the game.[54]
  - Ben Abraham, a world-leading climate researcher focusing on the environmental impact of digital games we interviewed, often supports teams in tracking their climate impact.[55]

- **Be realistic about your capacity.** Honestly accounting for your capacity and resources is also a critical part of practicing your values. If a social movement game is created under discriminatory or inequitable working conditions, then that complicates the impact of the game. Keeping a game simple can reduce crunch.

Participatory Design and Community Inclusion

Your collaborations may only extend to the folks in the team you're working with but, for many social movement games, it is important to reach out and engage with broader communities. Collaborating with your intended audience can also be a productive way of creating social movements *through the process* of designing a game, as we talked about with *Never Alone*.

- **Create a discord community** while you're still developing the game so you can understand your audience, the community around your project. This can be valuable, despite the labor and time required. Community manager roles are increasingly necessary although, unfortunately, this labor is often put on women and otherwise marginalized folks.
- Even if you don't have a Discord community, find digital and non-digital ways to **get feedback throughout the process** from the communities you're working with. **Playtest and iteratively develop** with their feedback. To save costs and effort, start with non-digital paper prototypes.
- Remember that **your players are your community** as well. As founder and studio director Karla Reyes (Anima Interactive, Square Enix, Code Coven) told us, "sometimes the players effectively become developers in a way, because they are contributing to the development experience."[56] Even after the release of the game, players find ways to redesign games, through modding, speedrunning, fanart, and more.
- Consider **multiplayer and cooperative games** where that is an option. *Never Alone* is a two-player cooperative game because they wanted to emphasize the game's theme of interdependence. (It's important to note that adding multiplayer into a game can greatly extend its production time.)
- **Collaborate with other organizations**, such as nonprofits. *Never Alone* was created with support from the CITC. *People Power: The Game of Civil Resistance* was created with assistance from the International Centre for Non-Violent Conflict.[57] The challenge of these collaborations is making sure you are aligned on values and that the expectations of the organization fit with your team's capacity.
- **Design *with* communities, not *for* them.** Review toolkits for collaborating with specific cultural communities, such as On-Screen Protocols and Pathways.[58] Consider which stories make sense for you or your team to tell. Make sure you're not just creating more work for people, and try to compensate community members fairly for sharing their knowledge and experience.
- **Join communities of folks doing similar work**. Find like-minded creators and share resources with them. Join competitions like Green Game Jams to experiment with creation. Just the act of gathering and creating in community can be a radical opening for further collective action. Having a community of like-minded people and sharing resources with each other can also be a great way of preventing burnout.

- **Tell stories collaboratively** and use storytelling to keep archives of your work for generations to come. This can be done through devlogs, blog posts, social media posts, internal communications, and many other ways. In our interview with Ivan Marovic from the International Center of Non-violent Conflict (*People Power: The Game of Civil Resistance*), he emphasized that the "essence of a (game) is a good story."[59]

Simply the process of engaging with communities can, if done right, be deeply impactful. That said, not all interventions in communities are beneficial if not based on expressed needs. As Nicklin explained in our interview, sometimes the most helpful interventions in community or organizations are to just donate or go "to your local coop (to do the) washing up."[60] Narrative designer and producer Pablo Quarta of Matajuegos recommended: "try to collaborate with activism that is already happening by going to the communities involved and asking 'how can we, as game designers, help out and bring something to the table?'"[61]

CONCLUSION

The move toward collaboration and equitable labor is the mycelial network of change-making that will produce the mushrooms we rely upon.

Through collective action, we can share the workload and each one of us can be a small but significant part of a larger, collaborative movement. Don't worry about doing everything in this toolkit; focus on making things that match your circumstances, whereby you can collaborate with others and see the impact of your efforts.

You also don't have to know everything. Sometimes the most effective way forward is to open things up, to encourage the player to explore new questions and ideas without dictating the answers. As Jayanth said in our interview:

> The key is to leave the player with more questions than answers, and then: 'go look, the real world is available for you. Go and go and answer some of these things.'[62]

NOTES

1 Alan Gershenfeld, Zoom Interview, August 23, 2024.
2 Upper One Games. Never Alone. E-Line Media. PC, PS3/4, Wii U, Xbox One, and Nintendo Switch. 2014.
3 Gershenfeld.
4 E-Line Media, "Never Alone: A 10th Anniversary Retrospective," *Medium* (blog), November 19, 2024, https://medium.com/@elinemedia/never-alone-a-10th-anniversary-retrospective-d5d69bd7d0ca.
5 Aric McBay, *Full Spectrum Resistance: Building Movements and Fighting to Win*, vol. 1 (New York: Seven Stories Press, 2019); Aric McBay, *Full Spectrum Resistance: Actions and Strategies for Change*, vol. 2 (New York: Seven Stories Press, 2019).

6 "Casual Games for Protesters," accessed December 9, 2024, https://www.protestgames.org/.
7 Inkle. 80 Days. Inkle. PC, Nintendo Switch. 2014
8 Failbetter Games. Sunless Sea. Failbetter Games. PC, PS4, Xbox One, and Nintendo Switch. 2015.
9 Meghna Jayanth, Zoom Interview, September 3, 2024.
10 Jayanth. Jayanth credited the term "rehearsing the future" to Ruth Wilson Gilmore's work on abolition as practice. UCL, "Transcript: In Conversation with Ruth Wilson Gilmore," Sarah Parker Remond Centre, September 11, 2023, https://www.ucl.ac.uk/racism-racialisation/transcript-conversation-ruth-wilson-gilmore. This idea is also explored in Robin Maynard and Leanne Betasamosake Simpson's book *Rehearsals for Living*, which reflects on and practices how we "build livable lives together in the wreckage." Robin Maynard and Leanne Betasamosake Simpson, *Rehearsals for Living* (New York: Alfred A. Knopf Canada, 2022).
11 Rosa Schwartzburg, "The US Military Is Embedded in the Gaming World. Its Target: Teen Recruits," *The Guardian*, February 14, 2024, sec. US news, https://www.theguardian.com/us-news/2024/feb/14/us-military-recruiting-video-games-targeting-teenagers.
12 McBay, *Full Spectrum Resistance: Building Movements and Fighting to Win*; McBay, *Full Spectrum Resistance: Actions and Strategies for Change*.
13 Aric McBay and Pamela Cross, *Direct Action Works: A Legal Handbook for Civil Disobedience and Non-Violent Direct Action in Canada*, 2020.
14 Ensemble Studio. Age of Empires. Microsoft. PC. 1997.
15 Ensemble Studios, "Age of Empires" (Ensemble Studios, Xbox Game Studios, 2024 1997).
16 11 Bit Studios, *Frostpunk*. 11 Bit Studios. PC, PS4, and Xbox One. 2018.
17 Jennifer Estaris, Zoom Interview, September 6, 2024.
18 Joel Jordon, "Boss Battle," itch.io, 2017, https://phoenixup.itch.io/boss-battle; Joel Jordon, Zoom Interview, August 22, 2024.
19 adrienne maree brown, *Emergent Strategy: Shaping Change, Changing Worlds*, Illustrated edition (Chico, CA: AK Press, 2017).
20 "Palestine Skating Game," itch.io, 2024, https://palestineskatinggame.itch.io/prototype3.
21 E-Line Media. 2014.
22 Matajuegos. Atuel. Matajuegos. PC. 2022. https://matajuegos.itch.io/atuel.
23 Molleindustria. Oiligarchy. Molleindustria. PC. 2008. https://www.molleindustria.org/en/oiligarchy/.
24 Littleloud. Sweatshop. Littleloud, Channel Four Television. PC. 2011.
25 Die Gute Fabrik. Saltsea Chronicles. Die Gute Fabrik. PC, PS5, Nintendo Switch. 2023.
26 York Zimmerman, Inc. People Power: The Game of Civil Resistance. International Center on Nonviolent Conflict. PC. 2010.
27 Matajuegos, "Atuel."
28 Matajuegos, "Union Drive," 2021, https://uniondrive.itch.io/union-drive.
29 "Save Amazonia Minecraft Project in Brazil (Video Grab)," Greenpeace, 2023, https://media.greenpeace.org/archive/Save-Amazonia-Minecraft-Project-in-Brazil--Video-Grab--27MZIFJLL48AV.html.
30 Mojang Studios. Minecraft. Mojang Studios. PC. 2011.
31 Nintendo. Animal Crossing: New Horizons. Nintendo. Nintendo Switch. 2020.
32 Maxis. The Sims 4. Electronic Arts. PC, PS4, Xbox One. 2016.
33 Rockstar North. Grand Theft Auto V. Rockstar Games. PC, PS4, PS5, Xbox 360, Xbox One, Xbox X/S. 2013.
34 Mojang Studios, "Minecraft."
35 "Towards a Free Palestine," itch.io, 2024, https://itch.io/jam/towards-a-free-palestine.
36 "Games for Change," Games for Change, accessed December 9, 2024, https://www.gamesforchange.org/.

37 Joel Jordon, "Practical Aesthetics: How to Use Art as Argument – My Different Games Conference 2018 Talk," *The Game Manifesto* (blog), October 15, 2018, https://gamemanifesto. net/2018/10/14/practical-aesthetics-how-to-use-art-as-argument-my-different-games-conference-2018-talk/.

38 Hannah Nicklin, Zoom Interview, September 6, 2024.

39 Kara Stone, Zoom Interview, September 5, 2024.

40 Estaris, Zoom Interview.

41 "Never Alone (Kisima Ingitchuna)."

42 Paolo Pedercini, Zoom Interview, August 20, 2024.

43 Stone, Zoom Interview; Kara Stone, "Solar Server," Solar Server, accessed December 9, 2024, https://www.solarserver.games.

44 Nicklin, Zoom Interview.

45 Ben Abraham, Zoom Interview, August 27, 2024.

46 Ustwo Games. Monument Valley. Ustwo Games. PC, Android, Windows Phone. 2014.

47 Jordon, "Boss Battle."

48 Baby Ghosts, "What's the Difference between Studio and Project Funding?," Baby Ghosts, 2024, https://babyghosts.fund/news/studio-vs-project-funding.

49 Amy Fredeen and Alan Gershenfeld, "Gaming on a Mission: E-Line Media," *Exploring the Creative Economy*, 2018.

50 Baby Ghosts and Gamma Space, "Actionable Values," Weird Ghosts: Learn, accessed December 9, 2024, https://learn.weirdghosts.ca/studio-development/collectivism/actionable-values.

51 These resources were developed by Baby Ghosts (eileen mary holowka and Jennie Robinson Faber) with support from the Gamma Space 2023 Peer Accelerator facilitators, kaitlyn dougon, bryan dupuy, Henry Faber, and datejie cheko green. The website will be regularly updated as we continue to co-develop these resources with current peer facilitators.

52 Gershenfeld, Zoom Interview.

53 Nicole Carpenter, "Saltsea Chronicles' Creators Tracked the Game's Climate Impact to Create Industry Change," *Polygon* (blog), October 12, 2023, https://www.polygon. com/23914267/saltsea-chronicles-climate-impact-report-studio-die-gute-fabrik.

54 Gershenfeld, Zoom Interview.

55 Abraham, Zoom Interview.

56 Karla Reyes, Zoom Interview, August 27, 2024.

57 York Zimmerman Inc. 2010.

58 ImagineNative, "On-Screen Protocols & Pathways: A Media Production Guide to Working with First Nations, Métis and Inuit Communities, Cultures, Concepts and Stories" (Toronto, 2019), https://creativebc.com/wp-content/uploads/2022/02/imagineNATIVE_on_screen_ protocols_and_pathways_5_15_2019.pdf.

59 Ivan Marovic, Zoom Interview, September 3, 2024.

60 Nicklin, Zoom Interview.

61 Pablo Quarta, Zoom Interview, September 3, 2024.

62 Jayanth, Zoom Interview.

Together We Stand, Divided We Fall

7

How Can Game Developers Navigate the Economic, Social, and Environmental Sustainability Challenges of the Game Industry?

Thorsten Busch, Florence Chee,
and Tanja Sihvonen

ABSTRACT

In this chapter, we briefly explain why sustainability has to always be thought of holistically, i.e., across its three dimensions (economic, social, and environmental). We then discuss how these three dimensions matter to game developers. Lastly, we provide some high-level recommendations as to how game developers can improve their sustainability practices.

KEY TAKEAWAYS

- Sustainability is a complex concept that is about more than just trying to protect the environment. It addresses economic, social, and environmental aspects simultaneously.

DOI: 10.1201/9781003512400-8

- Game developers can benefit from identifying the specific economic, social, and environmental sustainability challenges in their industry. They will only be able to address them if they see them as a packaged deal.
- Game developers can improve their sustainability impact across all dimensions. We illustrate this with recommendations for individual game developers, game development studios, and industry associations.

INTRODUCTION

When it comes to dealing with sustainability, game developers are caught between a rock and a hard place. On the one hand, their industry is structurally unsustainable in many ways. On the other hand, game developers who want to do the right thing often find themselves working under exhausting conditions already, which makes meeting additional business demands or social and environmental engagement feel overwhelming or downright impossible. In this chapter, we will address these challenges to enable game developers to make more savvy decisions when it comes to managing their sustainability efforts. In Section 2, we will briefly provide some context and background to explain why sustainability issues should always be considered bearing their economic, social, and environmental dimensions in mind all at the same time. Section 3 will then illustrate what these three dimensions of sustainability mean for game developers. Finally, in Section 4, we will walk game developers through some points of intervention they can leverage, from the micro level (i.e., individual developers) to the meso level (i.e., organizational issues affecting studios) and all the way up to the macro level (i.e., the industrial structures that studios are embedded in). With this approach, we hope to provide game developers with a critical lens on the sustainability challenges of their industry and to enable them to actively engage with the individual, organizational, and systemic issues they wish to address in their work.

SUSTAINABILITY: WHY DOES ITS DEFINITION MATTER TO GAME DEVELOPERS?

Today, when lay people hear the term "sustainability," they usually think it means protecting the environment. When experts talk about it, however, they use a much more wide-ranging concept of the term. In order to walk game developers through the basics, we have to briefly acknowledge its historical and political background. The term originated in Germany as "Nachhaltigkeit" back in 1713, when it was used in forestry. The basic idea back then was that a kingdom should only log as many trees per year as would grow back the next year, so one would not run out of this important natural resource. Eventually, this idea made its way into international politics. In the 1980s, the United Nations in its *Brundtland Report* made this idea the guiding principle of international development. They defined sustainable development broadly as development that meets the needs of the present without compromising the ability of future generations to meet their own needs. So, while this definition includes an environmental dimension, it also addresses social and economic aspects. In the 21st century, then, the United Nations' *Agenda 2030* specified this broad idea of sustainability into 17 fairly specific *sustainable development goals* (SDGs). These are mainly directed at nation-states and centered around a wide range of economic, social, and environmental issues, but companies have been using them as guidance as well. For instance, the SDGs promote issues such as gender equality (SGD 5), decent work (SDG 8) as well as responsible consumption and production (SDG 12). So, when we discuss sustainability today, we do not merely refer to the natural environment. Instead, the concept of sustainability covers an enormous range of aspects that are all equally important. Much like human rights, one does not get to pick one aspect of sustainability over another. Instead, they all stand or fall together.

That is why we will discuss the sustainability challenges game developers face today along three classic dimensions of sustainability: *economic, social*, and *environmental*. This has become standard practice in sustainability work, not just in politics but also across a wide range of industries. Among business people, Elkington's *triple bottom line* (TBL) has been making a similar move since the late 1990s. The TBL asks businesses to not only care about their financial bottom line, but also to always consider their social and environmental bottom line at the same time. Elkington's terminology, along the lines of *profit, people*, and *planet,* essentially means the same things as the economic, social, and environmental dimensions of sustainability. In a bold and unusual move, Elkington issued a "recall" of his TBL in 2018 because he felt that many businesses over the past 20 years had been willfully misinterpreting it as being all about trade-offs between the three dimensions of profit, people, and planet. Elkington critiqued that many businesses had prioritized the economic dimension at the expense of their social and environmental impacts. We agree with Elkington's argument that this logic is harmful because it is at odds with the core of what sustainability is actually about. That is why, when businesses manage their impact on society, they should always keep the TBL in mind and simultaneously work toward all three sustainability dimensions—otherwise, they cannot achieve truly sustainable outcomes. Against this background, the following section will walk game developers through how the three dimensions of sustainability matter to their industry.

THE THREE DIMENSIONS OF SUSTAINABILITY IN THE GAME INDUSTRY

Since the big crash of the early 1980s, the video game industry has had an excellent run. The market kept growing from niche to mainstream as it constantly expanded to new consumer groups, and the industry got used to the idea that things would continue evolving in that same direction forever. The Covid-era boom in games proved that assumption right at least temporarily, but then the post-Covid crash with its tens of thousands of layoffs and stagnating investments made it abundantly clear that the game industry is not crisis-proof. This harsh turn of events raises the issue of economic sustainability.

Economic Sustainability

When taking a closer, long-term look at the economics underlying the game industry at large, it becomes obvious that the industry has not been evolving gradually and calmly since its conception. Instead, its evolution was often rapid, radical, and disruptive. And while it added many new market segments, platforms, and consumer groups, it also lost entire generations of game developers in the process and destroyed business models once thought to be future-proof. To an outside observer, this should come as no surprise because that is how volatile markets usually behave. But the post-COVID crisis shows that decision-makers at big game publishers overestimated the allegedly infinite growth potential of the industry they had gotten used to. They then over-corrected, and as a result, dozens of game development studios were closed and tens of thousands of game developers across the globe lost their jobs. The crisis has hit every type of studio, from triple-A to mid-sized studios and all the way down to small independent game developers. One major lesson from this is that game developers should not blindly trust markets to only ever go up. Instead, economic sustainability at both the industry and studio level is something one has to work hard for, and it also rests on factors outside the control of individual developers in many ways, such as global market and consumer trends.

Some positive consequences of the post-Covid game industry crisis include an increase in unionization efforts and the trend within the industry to critically question traditional business and ownership structures. Regarding unionization, increasing economic pressure on workers has brought attention to organizations such as *Game Workers Unite*, who are now garnering support previously thought impossible in the traditional industry climate of games. Regarding alternative forms of ownership, on the other hand, organizations such as cooperatives have garnered attention. The studio *KO-OP Mode* in Montréal, makers of *Goodbye Volcano High*,[1] have been a success story in this regard. Similarly, Swiss indie game developer *Stray Fawn Studio* has had remarkably sustainable success using an egalitarian model despite the fact that from a mere cost standpoint, Switzerland is a terrible location for making games. These examples show that when it comes to economic sustainability, the industry could benefit from more diversity in terms of business models. However, strategies that work for small players may not work as easily for mid-sized studios or mainstream publishers, so one cannot recommend any "one size fits all" approach in this context. One thing that holds true for everyone, however, is that game developers need economic sustainability if they want to have the resources and mental bandwidth to address the other two dimensions of sustainability.

Social Sustainability

When it comes to social sustainability in the game industry, two issues have been especially persistent: toxic workplace cultures, where "crunch" and "bro culture" stand out as particularly unsustainable, and toxic gamers, i.e., player communities that have been organizing on social media and within multiplayer games in order to openly and aggressively oppose or undermine the game industry's efforts at promoting diversity, equity, and inclusion (DEI).

When it comes to unsustainable workplace practices, crunch has been an issue across the industry for decades now, and we do not need to lecture game developers about it. It is widely known among developers that crunch often ends up producing poor outcomes in terms of product quality. Nonetheless, it has been normalized across the industry for decades by managers and producers making economically unsustainable short-term business decisions for understandable but ultimately self-defeating reasons. So, crunch is where economic and social sustainability connect because developers pay the price for decisions publishers make under the conditions of disruptive and unsustainable global market conditions. No amount of the oft-cited "passion" on the part of game developers will fix this problem. Instead, it needs to be acknowledged and managed proactively, and studios need to be enabled to create a healthy workplace culture that works in their respective contexts.

The second social sustainability issue we want to focus on briefly is the "bro culture" that has been normalized for decades across game cultures, development studios, publishers, and the games media. On one hand, this includes internal processes at publishers and game development studios, where numerous reports and court cases over the years have made it abundantly clear that many game developers have to deal with the consequences of toxic and unsafe workplaces. Prominent court cases in this context have involved some of the largest game publishers on the planet, such as *Activision Blizzard* and *Ubisoft*, as well as high-profile studios such as *Riot Games*. On the other hand, we should point out that this type of toxic culture does not happen in a vacuum. Instead, it is embedded in a social context in which games have been the object and trigger point for culture wars for decades now, but especially since the infamous #*Gamergate* campaign back in 2014. In terms of social sustainability, the issue game developers have had to deal with in this context is toxic gamer culture, and the game industry at large has struggled to come to grips with its obligations regarding DEI for many years. Toxic gamer culture becomes particularly visible whenever a large, orchestrated online hate campaign against game developers gains traction online, but it also involves less public practices such as gatekeeping among online gaming communities competing for cultural power. In recent years, this toxic culture has expanded into gameplay streaming, cosplay, and other popular fan practices, creating an environment in which game developers often feel unsafe and outright threatened.

Summing up the most pressing social sustainability issues in games, it becomes clear that how folks in and around the game industry manage the boundaries associated with work, rest, and play is complicated. Frequently, game developers get squeezed into contexts where they are trapped between unacceptable behavior from both the top of their own companies and extremists on the outside. Keeping in mind the connection between the three dimensions of sustainability is important in this context because economic pressures make it difficult to improve the social sustainability game developers deserve. At the same time, game developers working under terrible working conditions will not enjoy the freedom and mental bandwidth to address the environmental aspects of their work.

Environmental Sustainability

When it comes to the environmental dimension of sustainability, the game industry has a lot in common with other tech industries. Put bluntly, the industry's value chain is entirely dependent on energy-hungry hardware and an unsustainable use of natural resources.

Whether they are aware of it or not, and whether they want to or not, game developers have an impact on our natural environment at every step of the game-making process. While forward-thinking game developers might hope that clean and sustainable energy sources as well as more power-efficient hardware will fix this unpleasant aspect of their working conditions, the reality is that energy consumption and greenhouse gas emissions have been going up instead of down, and they will likely continue to do so in the future. That is because, among other factors, recent game design trends around cryptogames, NFTs, blockchain integrations, and similar energy-consuming applications have been exacerbating the negative environmental impact of the game industry even further.

While many big publishers, such as *Activision Blizzard*, *EA*, and *Ubisoft*, have been addressing environmental aspects of their business publicly via sustainability reports in recent years, one might wonder whether they are merely engaging in green-washing efforts. All things considered. Sure, their corporate responsibility and sustainability managers at the operational level will likely be forward-thinking individuals who take sustainability to heart. On the other hand, however, when interviewed by the business press, top managers at many big publishers have indicated that their priority is to invest in AI and other technologies aimed at making their organizations more efficient and productive. That is understandable given the current hype and FOMO (fear of missing out) around AI across many industries, but it also invites a healthy dose of skepticism on the part of game developers when it comes to the priorities of their top managers.

On a positive note, scholars and activists in recent years have found playful ways of dealing with the environmental impact of games. For instance, under the umbrella of the *United Nations Environment Programme*, a multi-stakeholder network has started to address this issue collectively by promoting the "Playing for the Planet" program. Moreover, the *IGDA Climate Special Interest Group* has identified resources that help game developers to improve the current situation. When it comes to trying to find out what their overall climate impact is, individual game studios can use tools such as life-cycle assessments of their entire product lines and then adjust their business models toward environmental sustainability wherever possible. Yet again, it is important to note how the three dimensions of sustainability mutually depend on one another, because game developers who work in a precarious market and who are exhausted because of toxic work culture will not have the awareness and mental bandwidth to be able to address their environmental impact. In the next section, we will make a few high-level suggestions as to how game developers can do their important jobs in ways that are more sustainable across every dimension.

INTERVENTION POINTS: HOW CAN DEVELOPERS BECOME MORE SUSTAINABLE?

In the following section, we will show some points of intervention for game makers, from very small things that apply to individual game developers (i.e., the micro level of analysis), to organizational practices for groups of game developers, such as a development studio (meso level), and all the way up to the big industry structures that studios find themselves embedded in both nationally and internationally (macro level).

The Micro Level: Individual Game Developers

When it comes to the things that literally every single game developer can do to imme-diately improve the sustainability of the game industry in small but relevant ways, the first step, as with so many things, is to educate oneself. Reading this book is an excel-lent start, and then getting in touch with other like-minded developers to build alliances and become sustainability ambassadors inside their respective companies is even better. Reaching out to others is vital because it is very difficult for individual developers to perceive sustainability challenges when they are toiling alone in the hamster wheel, and it is even more difficult to achieve any significant change without the support of allies. For instance, if developers are embedded in crunch culture for a long time, crunch becomes so normalized that individual game developers have a hard time recognizing it as such. And if one does not recognize a problem in the first place, one will not be able to fix it.

That is why honing one's critical reasoning skills and one's perception of unsus-tainable behaviors is the first step toward meaningful change, and only then will better behaviors and habits follow. For example, at some studios, colleagues will encourage each other to go home after 5 pm, whereas at other studios, constant pressure to stay longer is the norm. Whenever it is appropriate and reasonable, we would encourage game developers to be the kind of person who reminds their colleagues that it is OK to go home after a long work day, and we say that as academics whose work environment tends to be rife with crunch as well, so the irony is not lost on us.

Finding people who share an interest in games and sustainability has gotten easier in recent years thanks to initiatives under the umbrella of political organizations, such as Playing for the Planet, as well as developer-led initiatives such as the *IGDA Climate Special Interest Group*. The developers of the 2024 Rogue-lite "Ultros,"[2] to name just one positive example, thank the *IGDA Climate SIG* in the credits of their game, and in its graphics resolution options, the developers let players know about the energy impact of playing at different resolutions and frame rates. That is an elegant and admirable way of educating players about the environmental impact of their gaming choices and habits without sounding preachy or being overly wordy about it.

This also points to another obvious but difficult thing game developers can do, which is to leverage their unique skills and passion to design the content, stories, and mechanics of their games in ways that address sustainability issues in a compelling manner. Depending on context, game design, and audience, this can be more or less easy to do, and to be honest, there is always the danger of coming off as too preachy, obvious, or obnoxious. When done with finesse, however, this is an excellent opportu-nity to entertain players and educate them at the same time. Again, seeking feedback from developer peers and sharing best practices is essential in order to get it right.

The Meso Level: Game Studios and Organizations

When it comes to running a game development studio and managing the inner workings of an organization, game developers can find collective ways to support each other. Most importantly, game studios have two main levers to pull when it comes to improving their sustainability: structure and culture. With regard to structure, organizations can be run on different kinds of hierarchies and decision-making processes. Some of those might be flat, inclusive, or democratic, others might be more centralized. Moreover, some organizations are unionized, which has consequences for the structure and perceived legitimacy of decision-making processes. Ownership structures also play an important role, as companies might be owned by a single person, a group of investors, or by their employees themselves (as is the case with a cooperative), for instance.

Stakeholder relationships are also an important part of the structure that organizations are embedded in. By definition, a stakeholder is anyone who can affect, or is affected by, an organization's business activities. This includes a wide range of actors, such as customers, employees, or investors, for instance. A well-run company should know who its stakeholders are and put communication channels in place that allow these stakeholders to voice their needs and concerns. Those concerns will often be in conflict with one another, so managing stakeholder groups is a permanent challenge for organizations. Generally speaking, stakeholder groups will appreciate an organization's efforts to include them, and organizations are well-advised to always take into account their stakeholders' needs to the best of their ability. Typical tools for addressing this in practice include holding regular meetings with all stakeholder groups as well as running a materiality analysis to identify the most pressing issues that affect stakeholders and the company itself.

When it comes to developing an organizational culture centered around sustainability, it helps if companies have a clear mission and vision aside from just wanting to make money. A mission statement answers the question, "What are we here for?" A vision, on the other hand, describes a worthy cause and better future that an organization wants to achieve in the next five to ten years. If a company does not know what it is here for, chances are it also has no idea how to treat its members in ways that are more than just instrumental, transactional, or even disinterested. This is also where a company should ask itself whether it sees the world as a set of resources to exploit for its own gain, or whether it is willing to use its resources to help the world overcome its many sustainability challenges.

In order to make it easier to support each other and do the right thing within an organization, companies often give themselves guidelines, such as a code of conduct. Such guidelines should be co-developed with employees and not just dictated to them by management. Ideally, this will encourage and enable a culture of integrity, where employees want to do the right thing for the right reasons instead of being guided by formalistic rules and threats of disciplinary action in case they do the "wrong" thing. Let us be clear: establishing and developing a long-term sustainability culture within an organization that covers all three dimensions of sustainability in an inclusive fashion is one of the hardest ongoing challenges to figure out for any company. Nonetheless, if game developers wish to achieve a more sustainable future, they will need to argue for it within their organizations and establish alliances and networks both internally and externally. That brings us to the final aspect game developers need to take into account.

The Macro Level: Industry Structures and Global Markets

The macro level refers to any governance structure and institution that is bigger than an individual company or organization. Obviously, all game developers are in some way affected by forces far beyond their control, such as global markets for resources, technologies, labors, and consumers. This can feel daunting and disempowering when it comes to addressing sustainability challenges. However, while at least being honest, just shrugging one's shoulders and cynically stating, "Well, that's capitalism for you, there's nothing to be done" does not improve anything for anyone, so it is not a valid option.

Instead, game developers can join forces to push for change. For instance, they could try to lobby for industry-wide labor organizing. This will be challenging to implement both locally and globally. In the short-term, game development studios can become members of a wide range of networks, initiatives, and institutions centered around corporate responsibility and sustainability. For example, the *United Nations Global Compact* is a network of more than 25000 businesses that have committed themselves to respecting and promoting a set of ten principles centered around human rights, fair labor, environmental engagement, and anti-corruption. Another promising example of a sustainability initiative, in this case with environmental sustainability at the center of the operation, is the newly established *Sustainable Games Alliance*, a cooperative whose members are game development studios. Historically, self-regulation initiatives such as PEGI and the ESRB have demonstrated that the game industry can kick into high gear and get organized when it needs to. Those initiatives were reactive instead of being proactive, however. In order to shift the game industry as a whole toward more sustainable structures and ethical practices, we will need a critical mass of individuals to act in our collective interest. This may involve reaching across sectors to include the triple helix of academic, governmental, and industrial actors.

CONCLUSION

In this chapter, we have given an overview of the many challenges the game industry is facing today across the economic, social, and environmental dimensions of sustainability. Many of these challenges are structural and thus difficult to address by individual game developers alone. Nonetheless, we have made some general suggestions as to how game developers can try to do their part when it comes to making the game industry more sustainable. Ultimately, it is game developers themselves who will need to form alliances to find solutions, not academics. We do hope, though, that we have given game developers some food for thought and a framework through which they can identify the challenges they wish to address.

NOTES

1 KO_OP. Goodbye Volcano High. KO_OP. PC, PS 4, and PS 5. 2023.
2 Hadoque. Ultros. Kepler Interactive. PC, PS4/5, Xbox Series, and Nintendo Switch. 2024.

Making Every Game Green

8

Arnaud Fayolle

ABSTRACT

This chapter advocates the idea that every game can have an effective pro-socio-environmental impact, but each in its own unique way. It demonstrates how to identify a game's "Unique Impact Potential" by connecting its themes and gameplay and target audience to environmental and social considerations. Each section comes with practical question sheets to help developers brainstorm collectively and identify what the Unique Impact Potential of their game is.

KEY TAKEAWAYS

- A framework to identify the Unique Impact Potential of your game, based on its themes and gameplay, and how to articulate it to engage your specific target audience
- A definition of the four predictors of pro-environmental behaviors, to maximize the transformative potential of your games

MAKING EVERY GAME GREEN

"We made video games!"

That's what you and I will say when our grandchildren ask, "what did you do when you learned about the climate crisis?"

This answer may seem paradoxical today because we all know the terrible environmental impact of the digital industry (see Chapter 6 by Busch et al.). Yet, most developers I've met happen to be well aware of the potential of games to inspire cultural and social change, slowly building the critical mass we need for systemic change (more about that in Chapter 3 by T. York and C. Whittle). A truly herculean task, but absolutely worth trying!

DOI: 10.1201/9781003512400-9

Toward the end of last decade, I shifted from a long career of Art Direction in AAA game production to help developers make their games more sustainable and increase their pro-environmental impact. I co-founded the IGDA Climate Special Interest Group,[1] ran creative workshops for the Playing for The Planet Alliance,[2] and am now Game Sustainability Project Director at Ubisoft where I advise a dozen studios worldwide.

Most game developers I meet are excited about this idea, yet inevitably ask the following questions:

- **How could my current game ever be green?** "My game genre and story aren't about the environment, and I don't want to displease my players with something they aren't looking for!"
- **How can we know it will work?** "We can't measure if players really behave sustainably outside of the game, so we could be fooling ourselves!"

This chapter will explore these questions and provide evidence, references, and practical tools so that you and your team can find your own answers to move forward with confidence. Let's start by answering the second one, as it will inform the right way of thinking about the first one.

How to Make Green Games Effective?

As game developers, our time is precious. We won't invest in anything unless we're confident it will bring value to our players and increase our KPIs, whether they're about extending the game's reach, engagement, or social impact. To measure social impact (such as the success of a pro-environmental activation), we have tools to measure impact in-game, like metric collections or pre-/post-gameplay surveys. Unfortunately, these metrics only tell us about a player's intentions, not their actual behavior after they're done playing, and we're all aware that there can be a gap between intentions and actions (see Chapter 3).

To help bridge this gap, the IGDA Climate SIG compiled research to identify key **Predictors of Pro-Environmental Behaviors.**[3]

It appears that, while acknowledging that players might have physical or social barriers in their lives that might prevent action, they can be encouraged to adopt pro-environmental behaviors when supported with:

- **Knowledge**, because they can't act if they don't know what's happening. Knowledge ranges from awareness about a problem to understanding its components and the systemic aspects behind why it happens and how to take direct action.
- An **attitude** that inspires action, because they won't act if they don't care. The player's attitude refers to how they define themselves and feel about their relationship with the natural world. Each experience makes them re-examine their attitudes as they learn through emotional interactions, exposure to social or conceptual norms, and practice positive behaviors.
- A **perception of self-efficacy**, because they won't act if they don't think it matters. Players can build it by witnessing the effect of their in-game actions

on the virtual world, so they feel real and valuable, starting with small actions and gradually tackling bigger and more complex ones, and experiencing what the collective power of community action can achieve.

- And finally, they need **hope** that all of this is worth doing! Hope is critical for taking the first step and to keep going in the long run. It can be built by experiencing and practicing trust in other players, in their agency to set their own goals, and their ability to experience success.

This means that for our game to be effective in nudging players to act more sustainably, it needs to share specific knowledge about the problem, create an attitude-shaping experience that makes players want to act on it, while giving them the right tools do something in this direction and experience success! These predictors are powerful indicators to identify what elements need to be featured in our game, as mechanics, aesthetic, or narrative elements, to achieve our transformational objectives.

Now that we understand what the goal is for our design, let's return to question number one, "How could my current game ever be green?"

How to Make Any Game Green?

The willingness to do good can be easily extinguished by the fear of losing players' engagement, as developers often mistakenly assume that the genre or story of their games won't be suitable for pro-environmental messaging, or that their current player bases wouldn't want to engage with such content. Imagine telling your average First Person Shooter player: "Hey, do you know that biodiversity is collapsing? What about we stop shooting virtual people and plant trees instead?" I admit they probably won't be very receptive.

But what makes a game "green" isn't its genre or theme: it's the expectation that playing it will somehow push the needle toward a more sustainable world (more about this in David ten Cate chapter). Based on the extraordinary creativity, I've witnessed among development teams so far, I genuinely believe that every game can indeed become a force for positive change, regardless of their themes, genre, and player base. The catch is that no single recipe will ever work in all contexts, so it can seem tricky to figure out what will work with any specific game.

A more successful framing might be: "Hey, you're a brave warrior, and you're ready to defend what's yours. Your Home is being attacked: are you going to fight back?" You can imagine how this might be more attention-grabbing.

To determine efficiently and confidently what kind of pro-socio-environmental content would fit and enrich our games, and increase engagement from our current player base, we need to identify what makes our game unique.

THE UNIQUE IMPACT POTENTIAL

We game makers always dedicate time to define our game's Unique Selling Points (USP), to identify what makes our game stand out among competitors and please our target audience. Having clear USPs is crucial both for aligning the team behind a shared vision and selling the experience to our players, but if our objective is to inspire and engage players toward pro-environmental goals, this is not enough. To maximize our chances of positive impact, we also need to define our game's Unique Impact Potential (UIP): what cultural transformation is our game uniquely suited to achieve?

We can triangulate our UIP by looking at:

- The game's **Themes**: what it is about, and what is the context of its world and characters?
- Its **Gameplay**: what do we ask players to do in-game, and how can they solve problems?
- And of course, its **Audience**: who will play the game, and what is the context of their lives?

We'll look at each of these concepts in detail, then provide a bunch of practical questions to brainstorm with your team and adapt them to your specific game.

Our Game's Themes

There are games about everything!

Most games have a few key themes, often identified early during a conception phase, and reinforced throughout all stages of production to the point they even end up mentioned on the game's box or store description. But games can also have several secondary themes, used to define the game world, some characters' arcs, or specific moments. When picking themes, game makers are often concerned about approaching serious topics such as climate change, as they can't see how this would relate to their other existing themes. However, climate change is a very systemic topic that goes much further than just managing our greenhouse gas emissions; the causes are deeply rooted in everything that makes our modern civilization, and its consequences affect everything and everyone on the planet!

While this notion can be scary, it highlights that no matter what our game is about, it can be connected to some aspects of the climate crisis and its associated socio-environment goals. Anything related to human activities or natural phenomena can be a doorway to make a meaningful statement, showcase a character taking actions, highlight a nugget of science, or visualize an environmental problem or solution! *The Legend of Zelda* demonstrated this years ago by showing how an unsupervised mining operation can turn a biodiversity hotspot into a barren desert in *Skyward Sword*[4]!

One way to identify how our themes could relate to key aspects of our environmental crises is to look at the 17 Sustainable Development Goals (SDGs)[5] defined by the United Nations Environment Program to help humanity make sense of this complex mess and restore balance of the planet. Working with the 17 SDGs has the advantage of access to a wealth of documentation, guidelines, and advice from trusted sources, which adds legitimacy to a project. When participating in annual Green Game Jams[6] organized by the Playing for The Planet Alliance (facilitated by UNEP), our teams at Ubisoft discovered many links between their games and the 17 SDGs:

Games with very specific themes were great fits for specific issues. SDG "#14 Life Below Water" was a no-brainer for *Hungry Shark Evolution*[7] and *Hungry Shark World*,[8] and both game teams quickly produced content about protecting endangered marine species that naturally fit and enriched their game, generating strong engagement from their players.

Broader games which offered more freedom with open worlds, character creation, and non-linear campaign systems could pick from a larger range of relevant goals: *Riders Republic*[9] could relate to #3 Health & Well-being, #5 Gender Equality, #11 Sustainable Communities, or #15 Life On Land and make meaningful content that would match the game's DNA from all of them. They finally chose #13 Climate Action and used their community-building mechanics to organize the first in-game Climate March[10] in history. This made the community engagement visible to everyone, leveraging social normative learning before pointing at a potential solution—an in-game reforestation activity—to create a perception of self-efficacy among players.

The more systemic a game is, the more themes it can potentially address. City-builders like *Anno 1800*[11] let players experience how all these different themes are interconnected in our civilization, and thus almost all 17 SDGs were meaningful to explore! They ended up producing a game scenario called *Eden Burning*[12] where players are tasked to get rid of polluting infrastructures and build a dam as a green energy source all while taking care about the environment of the island (a combination of SDGs #6, #7, #8, #9, #11, #13, #14, and #15). This scenario quickly became a fan-favorite and triggered many online discussions about how to tackle this complex situation.

Whatever potential environmental themes we end up identifying for our games, it's important that they resonate with the team that's going to build the content.

Team Inclination

Teams that get passionate about certain pro-environmental themes would benefit from an intrinsic motivation boost and feel driven by a strong sense of purpose to dig deep into the matter, learn proper science, and tell relevant stories. Most teams participating in Green Game Jams have reported such experiences in their project post-mortems.

On the other hand, teams that don't resonate with such themes may perceive it as an unnecessary burden. They might keep to surface-level generalities and rely on approximations and stereotypes, which is understandable considering how hard game development is but ends up being both detrimental for the game and its social impact. If the team doesn't get excited about the first few socio-environmental themes we identified, just keep digging! As everything is related to the environment one way or another, we can always find more opportunities to tell stories that are both relevant to our game and meaningful to the team.

The best way to get everyone's interest is to engage them as early as possible in the ideation process. Across this chapter, you'll find a bunch of question pages you can use to brainstorm with your team. You'll be surprised how many potential pro-environmental cues can be discovered when using such collective intelligence, even in mainstream games!

TEAM PRACTICE FOLLOW BY BOX TITLE

IDENTIFYING YOUR THEMES' POTENTIAL

- What is your game about?
  - What is the purpose of the story?
    - What problem will players try to solve?
  - What is your game world?
    - How do its inhabitants impact their environment?
    - How does the environment impact its inhabitants?
    - Is this civilization sustainable?
      - What would it look like 10, 100, or 1,000 years from now?
      - What problem would arise?
      - How would they solve them?
  - Who are your characters?
    - What is their relationship with their environment?
    - How have they been impacted by it?
    - How do their activities impact it?
- How could your answers above be connected to current socio-environmental problems or solutions?
  - What SDG do they relate to?
  - Would surfacing them to your players enhance your game or distract from its initial goal?
- How much does the game team know about these themes?
  - Would they be excited to dig into them, or perceive them as a burden?
  - What socio-environmental matter are they passionate about?

Our Gameplay

Pro-socio-environmental themes can be found in any form of media, but let's face it: the reason our audience is playing games rather than watching TV or reading books is for our gameplay. And what a crucial transformative tool it is! Gameplay defines what the player's experience will be, dictating how and why they can interact with the virtual world around them. It informs what kind of solutions players are allowed to come up with to solve the problems the game will throw at them, and which attitudes will be rewarded or punished.

You might wonder what is the best type of gameplay to inspire pro-environmental action. I would say all of them, as tackling a systemic problem requires attacking it from all angles!

In-game mechanics mimic real-world actions, so let's look at what people do when they want to have a positive environmental impact in the real world. Even though we all agree that collective, systemic action is needed, people often disagree on what kind of action is the most relevant or urgent to start with.

Some claim that change should start with us: do our part, renew our connection with nature, learn to live with less, abandon our broken system and collectively build something new. In a word, learning to **Live Sustainably**.

Others say we should rely on our strong institutions, governments, and corporations to collectively transform our planet-crushing system into something that could sustain itself for centuries. In short: **Transforming the System.**

Some even say we should all be taking the streets right now, with banners and drums, holding our ground against those who destroy our planet for profit, with love and rage! In short: **Resisting against our planet's destruction.**

Who is right? Who should we trust? There are so many good people out there trying to convince other good people that their ways would be more efficient than the others, and the fear of advocating for the wrong solutions in our game can easily paralyze us. When trying to navigate these complex waters myself, I've collected about 40 different "climate actions" and mapped them into three action paths as seen on Figure 8.1. They seem to be going in different directions at first, but each one ends up bringing unique benefits to the other two, much like an ecosystem!

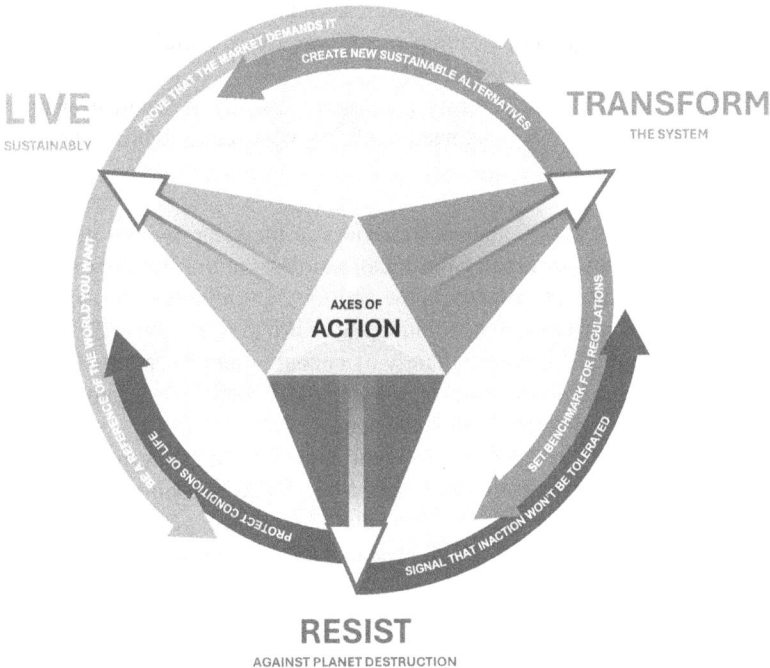

FIGURE 8.1   The climate action map. Graphic by the author.

Living sustainably helps those trying to transform the system by proving that there's a demand for it and acts as points of reference for those resisting the planet's destruction!

Transforming the system creates new sustainable alternatives for those who wish to live sustainably and sets new benchmarks for future regulations, allowing the resistance to take legal action.

Resisting against the planet's destruction sends strong signals to governments and corporations that inaction won't be tolerated, helping those working hard to transform the system from within, as well as protecting the conditions of life of those willing to live sustainably.

As always in nature, everything is connected: we can safely inspire actions in all three axes with our games and be confident that their outcomes will contribute to the grand tapestry of change. What does this mean for our gameplay, though? When considering our different game genres, it's easy to notice some natural potential to let players experiment with certain kinds of behaviors:

- Games about real-life activities, low-carbon sports, and relationships with animals or the natural world can inspire players to explore sustainable ways to live.
- Games featuring complex systems, such as city-building and tycoons are perfect to inspire societal transformation. Even games about high-carbon activities like car racing can set examples, the way *Trackmania*[13] does by featuring electric cars rather than petrol ones.
- Actions, combat, and shooting games can lead the resistance against our planet's destruction: we just need to clarify what we're fighting for or trying to protect!

Encouraging pro-socio-environmental behaviors in the same axis as our game genre will make our green features feel natural and seamless, in the continuation to the flow of the game that our players already know and love. Even better: we can then engage players in real-world activities that would feel like bringing the game's fantasy back to real life.! UsTwo Games did this brilliantly by encouraging players of *Alba: A Wildlife Adventure*[14] to take a series of real-world actions after playing the game because this would be what the game's hero would have done.

On the other hand, promoting pro-environmental gameplay in a different axis than the game's existing mechanics is a riskier bet. It may get ignored by some players or create unnecessary friction due to gameplay dissonance, yet it can also add some refreshing gameplay variety that could be welcomed by your most eco-conscious players (Figure 8.2).

TEAM PRACTICE

IDENTIFYING YOUR GAMEPLAY ENGAGEMENTS' POTENTIAL

- What does your game task players to do?
- What **action verbs** best describe your player experience?

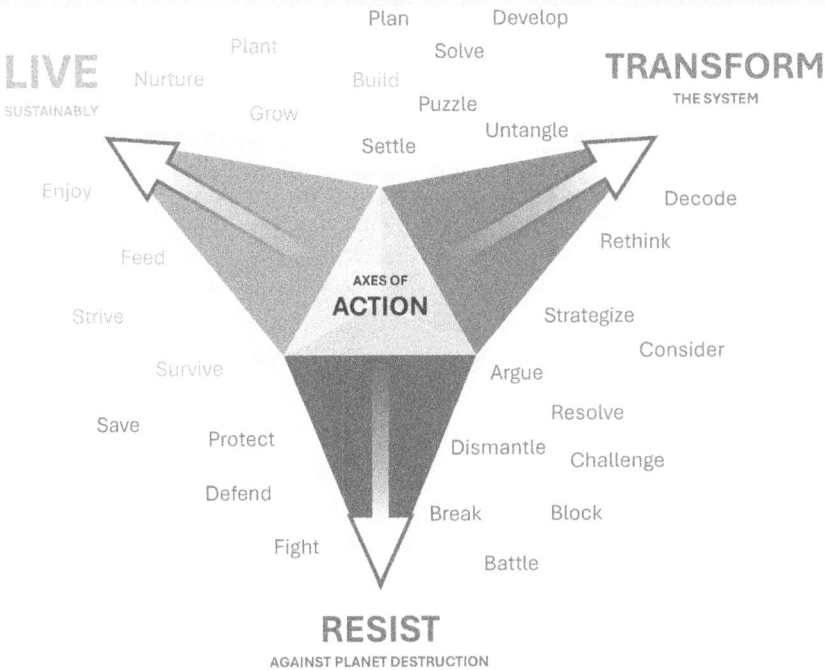

FIGURE 8.2   Using action verbs to navigate the climate action map. Graphic by the author.

- **Where** would you place your game on the Action Map?
- What gameplay mechanics toward **your themes** would fit this direction?
- What **real-world action** would you hope players want to take after playing your game? (be as specific as possible!)

Our Game's Audience

The last key piece of our puzzle is our audience: who are they? How are they playing and why? Many designers have learned the hard way that we are not our own audience, and what would work for us won't necessarily work for all those who will play our game.

Players' Personal Pro-Environmental Journey

As diligent game makers, we're doing the necessary research to know our audience. We have a good idea of their average age, gender, location, gaming habits, and maybe even some of their passions. We should also be interested in learning **where our audience stands on their pro-environmental journey:**

- Are they still ignoring the problem? If so, is it by accident or by choice?
- Are they already aware of the crisis, but aren't acting yet?
- Or are they already actively involved in making the world more sustainable?

Depending on our genre and themes, the proportion of each segment among our audience will vary greatly: we can imagine a niche game about foraging in the wild (like *Out & About*[15]) might be mostly played by actively involved nature-lovers. The proportion of eco-conscious players among our audience is likely to be affected by our existing themes and gameplay, while a highly mainstream game played by a billion players like *Subway Surfers*[16] may have a bigger proportion of unaware people, so we can adapt our speech to our audience. The proportion of eco-conscious players among our audience is likely to be affected by our existing themes and gameplay:

- Explicit environmental themes are more likely to attract players who already have a pro-environmental attitude.
- Radical gameplay explicitly aimed at restoring balance or questioning our current system might attract pro-environmental players but repel those who haven't progressed far enough in their journey yet.

Which player group our game should target is up to us:

- Targeting eco-conscious players allows us to dig deeper into environmental themes and gameplay, as we can expect they'll appreciate learning and discovering new things in this direction. It has greater transformational potential.
- Including unaware players in our target allows us to potentially reach a wider audience, but it might force us to be more subtle with our pro-environmental elements to avoid triggering adverse reactions.

This dilemma has pushed most AAA studios to be quite subtle with their pro-environmental messaging, as they need to reach wide audiences if they want any hope to make their expensive games profitable, yet exceptions exist: *Final Fantasy VII*,[17] which glorifies heroes taking arms against an extractive energy corporation to save their planet, was one of the fastest selling PS1 game (2 million sales in three days) and its 2022 remix one of the fastest-selling title of the PS4 (3,5 million sales in three days).

Yet, even players without prior interest in saving the planet may want to listen if we tell them about the personal benefits of sustainable behaviors instead: "Changing your diet is better for your own health! Using less power makes you save money!" Being good for the planet becomes a simple co-benefit.

Finally, if we have any doubt about our current audience's interest in environmentalism, it's even more crucial to just leverage fun and humor to subtly raise awareness and educate without being preachy. I was amazed by how countless deep pro-social statements and demonstrations *South Park: The Fractured but Whole*[18] managed to sneak into the game, leveraging absurd humor to make them entertaining. These contributions to society may be hard to measure, but I'm convinced they do help move the needle toward a more just, thus more sustainable world.

Players' Motivations

One last valuable approach is to consider our players' motivations: WHY are they playing our game?

If we can determine the reason why players came to play our specific game, we know what they expect to find in it. This can inform what kind of pro-socio-environmental content we can include, and how to present it. A good starting point can be to look at Quantic Foundry's Motivational Clusters[19] model. By using established psychometric methods and data from over 1.65 million gamers, they identified six major sources of motivations: Action, Social, Mastery, Achievement, Creativity, and Immersion. If we can identify the main motivations for players to play our game, we can deduce which activities can motivate them, both within and outside the game!

Imagine we want our game to tackle the problem of garbage collection and management. What kind of gameplay could inspire action?

- SOCIAL players may be interested in cleaning garbage for the sake of the community, like in *Alba: A Wildlife Adventure*.[20]
- Players from the MASTERY cluster would be more excited in how reusing can optimize both their playthrough and their lifestyle, like in *Under the Waves*[21] where collecting used oxygen cans let them build new ones for free.
- CREATIVE players would love to discover many creative ways to upcycle their garbage into pieces of art, like in *Yoshi's Crafted World*.[22]
- Players yearning for ACHIEVEMENT can find engrossing to go and find all the different kinds of discarded items in their game world and turn them into unique collectibles, like in *Bear & Breakfast*.[23]

Players after IMMERSION and ACTION would enjoy the intricate game world and/or fast paced gameplay of *Another Crab's Treasure*,[24] receiving knowledge and attitude-shifting stories about how our garbage affect marine life at the same time.

This illustrates the importance of considering our audience to identify the UIP of our game. Where our audience stands in their personal pro-environmental journey predicts how much they'll be willing to engage with explicit environmental themes and gameplay. Our audience motivations inform us about what kind of activities and experiences they'll be happier to engage with.

TEAM PRACTICE

IDENTIFYING OUR AUDIENCE'S EXPECTATIONS
- Where do we expect our audience to stand in their personal pro-environmental journey?
    - Are they still ignoring the problem? By accident or by choice?
    - Are they already aware of the crisis, but aren't acting yet?
    - Are they already actively involved in making the world more sustainable?

- Do we expect them to be **receptive** to **explicit** pro-environmental themes or gameplay?
    - Would hiding these themes broaden our audience?
    - What trade-off between transformation and reach would we be comfortable with?
- What **motivational cluster(s)** do we expect our players to belong to?
    - What kind of experiences would fulfill their expectations?
    - What do we want them to do after experiencing our game?

BRINGING EVERYTHING TOGETHER

Once we're clear about the potential of our themes, our gameplay, and our audience, it can be interesting to articulate our game's UIP in a way that can be remembered and shared to the wider team. It can even be presented to investors and stakeholders to demonstrate the game's potential for impact!

TEAM PRACTICE

FORMULATE YOUR UIP

Try using the following template, or create your own:
- Our game (name) is perfectly suited to inspire our audience (demographic) motivated by (motivation clusters) to engage in (axis of action) toward (environmental goal/SDG).

Therefore, it has a unique potential to activate players to: **(desired player transformation)**

CONCLUSION

I dream of a future where every time players finish a game session, they feel inspired, empowered, and equipped to make their world more sustainable. I am confident it can happen. We have seen how most game developers stood for the representation of women and minorities during the infamous "gamergate." We have seen them standing again to defend accessibility against a terribly angry portion of their player base who wanted to gatekeep their hobby for the most able and skillful only. Game developers, by and large, want to do what is right. It is not a lack of values, ambition, or community spirit that prevents us making every game green; but often just the subjective assumption that green games would not be palatable for our beloved players, and the sheer complexity of figuring out how to make them right!

I hope I helped you smash through these assumptions by proving that any game genre and theme can be relevant green game material if we weave them with our specific audience in mind. Perhaps you now have the tools to start a productive conversation with colleagues who might be so far still hesitating. Green games will always be complex beasts to tackle, as they stand at the intersection of two complex systems: videogames and environmental sciences. Yet, I hope the maps and compass in this chapter can help you navigate the sea of complexity with a clear destination in mind. Above all, I hope you now feel confident while making games that help players progress along their own pro-environmental journey.

So, thanks to you, who is holding this book, because you will be one of those making this dream a reality. So that, when our grandchildren ask us, "what did you do when you learned about the climate crisis?", we can proudly tell them:

"We made video games!"

NOTES

1 https://igda.org/sigs/climate/.
2 https://www.playing4theplanet.org/.
3 Whittle, C., York, T., Escuadra, P.A., Shonkwiler, G., Bille, H., Fayolle, A., McGregor, B., Hayes, S., Knight, F., Wills, A., Chang, A., & Fernández Galeote, D. (2022). *The Environmental Game Design Playbook (Presented by the IGDA Climate Special Interest Group)*. International Game Developers Association.
4 Nintendo EAD. *The Legend of Zelda: Skyward Sword*. Nintendo. Wii. 2011.
5 https://sdgs.un.org/goals.
6 https://www.playing4theplanet.org/green-game-jam.
7 Ubisoft London. *Hungry Shark Evolution*. Ubisoft. Android and IOS. 2015.
8 Ubisoft London. *Hungry Shark World*. Ubisoft. Android and IOS. 2015.
9 Ubisoft Annecy. Riders Republic. Ubisoft. PC, PS4, PS5, Stadia, Xbox One, and Xbox Series S/X. 2021.
10 Boris Maniora. "How Riders Republic mobilized its players for the first digital climate march." *Gameindustry.biz*. 2022. https://www.gamesindustry.biz/how-riders-republic-mobilized-its-players-for-the-first-digital-climate-march.
11 Ubisoft Blue Byte. *Anno 1800*. Ubisoft. PC, PS5, and Xbox Series X/S. 2019.
12 https://www.anno-union.com/devblog-eden-burning-green-game-jam/.
13 Ubisoft Nadeo. Trackmania. Ubisoft. PC, PS4, PS5, Xbox One, and Xbox Series X/S. 2020.
14 Alba, a Wildlife Adventure (UsTwo, Plug in Digital, 2020).
15 Out and About (Yaldi Games, Holysoft GmbH).
16 SYBO. *Subway Surfers*. SYBO. Mobile. 2012.
17 Square. *Final Fantasy XVII*. Square Enix, Square, Sony Computer Entertainment, and Eidos Interactive. PC, PS, PS4, Android, Nintendo Switch, and Xbox One. 1997.
18 South Park Digital Studios LLC, Ubisoft San Francisco. *South Park: The Fractured But Whole*. Ubisoft. PC, PS4, Xbox One, and Nintendo Switch. 2017.
19 Nick Yee & Nicolas Ducheneaut. *The Gamer Motivation Profile*. Quantic Foundry. https://quanticfoundry.com/gamer-motivation-model/.
20 UsTwo, 2020.

21 Parallel Studio. *Under The Waves*. Quantic Dream. PC, PS5, PS4, Xbox One, and Xbox Series X/S. 2023.
22 Good-Feel. *Yoshi Crafted World*. Nintendo. Nintendo Switch. 2019.
23 Gummy Cat. *Bear and Breakfast*. Armor Games. PC and Nintendo Switch. 2022.
24 AggroCrab. A*nother Crab's Treasure*. AggroCrab. PC, PS5, Xbox One, Nintendo Switch. 2024.

SECTION 2

Stories from a Sustainable Future

Section 2
Production, Processes, and Material Conditions

9

Clayton Whittle

In this second section, we move toward a more eclectic understanding of what it means to be a green game. We seek to challenge your understanding in this section, to provide you with broad and, on occasion, conflicting conceptualizations of what a green game or a green game industry means. We do this because we seek to elevate a critical self-examination of the green games movement, the processes that create green games, and the individuals or organizations that involve themselves in this movement.

In Chapters 10–13, we aim to expand the conversation of sustainability in games to include a holistic understanding of the role of both games and the game industry in our society. In expanding this definition, we open ourselves to more opportunities to reexamine and redefine.

CHAPTER 10: HOW CAN SUSTAINABLE GAME MAKERS GET JOURNALISTS TO PAY ATTENTION?

Writing from his perspective as a game industry journalist, David Lumb presents us with stories of individual creators and their green games. David's reporting here is gut-wrenching, as he does not shy from the truths of the potentially unsustainable human cost of games, the instability of employment, and the very real demands on any serious game industry journalist. David's chapter, at its heart, highlights a very real truth: it is not enough to do; one also must be seen doing. His work calls out the possibility of failure and reminds us that, however we define success—we must define it clearly or risk achieving nothing.

DOI: 10.1201/9781003512400-11

Too frequently, I have worked with studios that fail to define their desired impact. Through stories of hard-earned success, David reminds us that, if we cannot name our goal, we cannot pursue it.

There is no harm in simply creating. Art for the sake of art is a noble cause. But, in the world of green games and activism, we must ask ourselves what we wish to achieve by creating a green game and hold ourselves accountable for success. Without that accountability, we risk a subtle and dishonest greenwashing of our own experiences.

CHAPTER 11: BUILDING A GREEN GAME MOVEMENT: OVERVIEWING PRO-ENVIRONMENTAL EFFORTS IN GAMING

Valley Lopez unapologetically holds a mirror up to the people and the groups pursuing what he refers to as the *green games movement*. As the co-chair of one of the organizations examined in Valley's research, I find that his work directly and explicitly challenges the approaches I have personally and professionally advocated for.[1] Yet, this is a necessary challenge, and one that deserves to be amplified.

Through conversations with leaders across the green games movement, Valley works to illustrate that the movement lacks cohesion, coordination, and shared meaning. These real challenges are made apparent in the interviews conducted for this study, and the author presents a framework for realistic pursuit of a more organized, more aligned, and more effective movement.

In short, while the many actors in the space may be unique, Valley reminds us that they should not consider themselves disparate.

CHAPTER 12: EXPLORING SUSTAINABLE GAMEMAKING THROUGH AN ARTGAME RESIDENCY

The chapter written by Cindy Poremba, Kara Stone, Ian Garret, and Ben Abraham is perhaps one of our most unique. This team's work in designing and delivering a *renewable artgame residency* challenges our very notions of what it means to produce and develop a game. By working across geographic and cultural borders to provide creators with sustainable production environments, the residency discovers and reports inherent challenges and realities embedded in the production process.

In the pursuit of a more sustainable approach to the design of games, the residency outline in this chapter illustrates something that we must all recognize: our understanding of how to approach energy-conscious game creation is nascent at best. Without exploratory and experimental initiatives like this one, we cannot move forward in our understanding of what a truly green game industry really looks like.

CHAPTER 13: CHANGING THE QUESTION: WHAT IS AN ECOLOGICAL GAME?

Though at first glance, the final chapter in this section may appear to be so theoretical that it is removed from the real world of game development, David ten Cate's framework for understanding what it means to be an *ecogame* serves as a cumulative understanding of the lessons from the preceding chapters. David challenges us to see each game's ecological significance through the lens of the messages it communicates, the processes used to create it, and its role in our society as consumable artifact.

David's work is critical in that it allows us to see each of the messages of the preceding chapters as part of a whole, providing a framework through which we can examine and understand games from the societal experience, rather than the individual.

CHAPTER 14: GAME DESIGN IN CHINA: PIONEERING PRACTICES AND FUTURE DIRECTIONS

Vincenzo De Masi, Qinke Di, Siyi Li, and Yuhan Song offer maybe the furthest zoomed-out perspective on what systemic change could look like. By offering a critical discussion of how a system of cultural production that is different from the dominant Western one, the authors show how China and the Chinese game industry work toward sustainability goals. This may be a controversial argument, but there are a number of cases here that strongly suggest that representative democratic systems and open markets can learn some things about implementing sustainability requirements as top-down policy that requires compliance. There are downsides to this approach, not the least of which is democratic legitimation. However, we believe that having an open discussion about governance is a necessary part of intellectual honesty. One of the takeaways here is that building a strong policy like the Corporate Sustainability Reporting Directive that needs to be followed could be a meaningful step.

NOTE

1 Note that I individually participated in this project before this book was announced, and I recused myself from voting on the inclusion of this chapter.

How Can Sustainable Game Makers Get Journalists to Pay Attention?

10

David J Lumb

ABSTRACT

This chapter focuses on the intersection of sustainability games and media coverage. After an introduction, there is a brief section detailing the current state of games media, highlighting its challenges, which is intended to provide context for the reduction in quality and quantity of coverage at the time of writing. Then this chapter narrows its attention to case studies for sustainable game successes as a potential example to follow, as well as perspectives from games journalists. This chapter closes with advice for developers to better solicit coverage and what journalists should consider when covering sustainable games.

KEY TAKEAWAYS

- **The media gate is cracking.** Sustainability games struggle for coverage as game journalism collapses and algorithms dictate visibility.
- **Fun comes first.** Games like *Terra Nil* prove that engaging gameplay carries climate messages farther than education alone.
- **Pitch with purpose.** Strategic framing, real-world impact, and a sharp message are essential to break through the noise.
- **Build your own spotlight.** Community, storytelling, and social presence often outshine traditional press for indie developers.

DOI: 10.1201/9781003512400-12

In the last decade, there have been a couple of dozen articles in English-speaking mainstream games media dedicated to the video game industry and climate change. These articles slipped into the swirling deluge of daily gaming coverage—rumor reports, launch reviews, guides, best lists, and all manner of news—as drops in the proverbial bucket. However, nobly intentioned, eco-focused gaming articles push out against the wind of a conservative readership deaf to sobering information in their hunger for upcoming blockbuster games. These climate reports are published by a handful of true believers among a skeletal editorial industry desperate to sustain its bare-bones business.

The game industry itself isn't faring much better, with year upon year of historic mass layoffs, shuttered studios, and high-stakes launch expectations that seem doomed to disappoint. These devastations influence each other: fewer, bigger games carry the weight of studio futures, and a dwindling roster of media publications means less coverage, especially of smaller games with niche appeal. Gamers are robbed twofold, losing both new games and articles giving insightful context and recommendations. Influencers and streamers spill into the breach, most without journalistic training or ethics, largely dependent yet refusing to acknowledge the importance of rigorous journalism covering games and those who make them.

Despite this bitterly inhumane whittling of the video game business and experience, game developers have responded empathetically to the moment of global crisis and have started making games that engage with the calamity of our age. As creativity galvanizes despair into constructive action, developers use games as a lens for climate change to get players and the press to feel empowered. "Play is the antidote of doom," Trista Patterson, Director of Gaming Sustainability at Microsoft, told me in 2023.[1] So how do game developers—especially those in indie shops with few resources, as those on the cutting edge of sustainability gaming are today—get media attention to help spread stories about their climate change-aware games?

For developers fluent in the danger of climate change and distraught at the lack of coverage in the overlap between gaming and sustainability, I will explain what's straining games media's capability to do more than cling to its meager existence. Later on, I will highlight released and upcoming titles at that intersection, from the critically acclaimed and commercially successful *Terra Nil*[2] to the upcoming *Spilled!*[3] and *Outbound*.[4] Lastly, for developers creating similar sustainability games, I will also lay out best practices for connecting with media and journalists like myself.

A caveat up front: I am one of the journalists who has written about gaming and sustainability for a story published on CNET.com[5] in which I included the Patterson quote above (and quoted Patrick Prax, co-editor of this book). I speak from experience within the editorial industry, having written gaming articles for the past decade for a variety of publications. But I have also expanded this chapter to include perspectives from other gaming journalists and editors who parse through dozens of pitches every day.

A SHORT PRIMER ON TODAY'S
GAMING MEDIA

The internet and social media continually change how players connect with and learn about the games they enjoy, but it's more than market forces shifting how games are written about today. Over the last five years at the time of this chapter's writing in 2024, many prominent gaming-focused publications have been shuttered or suffered such extensive layoffs to effectively achieve the same, including Fanbyte (2022),[6] Vice Gaming/Waypoint (2023),[7] Washington Post's Launcher (2023),[8] Kotaku Australia (2024),[9] and the iconic Game Informer (2024).[10]

Game Informer's closure dealt a blow to games media's morale for ending a storied 33-year run as the last major print gaming publication on newsstands—but also for its suddenness, as parent company GameStop shut down the publication with no warning (what would have been the last issue was reportedly 70% complete). It's one of many examples of parent companies, including wealthy owners and venture capitalist overseers, making capricious decisions that gut and end publications on whims. On a lesser scale, this can lead to waves of layoffs that shrink newsrooms to meager staff who must do more with less.

Sadly, game developers know this dynamic all too well.

Sometimes, parent companies wringing their hands over layoffs and closures gesture vaguely to market conditions—and to some degree, it's accurate to say that games media has suffered the wider media calamity of shrinking income as ad revenue continues declining. Google and Facebook seized most of the advertising revenue of the modern internet, and games media play for scraps. In the last decade, online media sites have pivoted to affiliate marketing, earning bits of revenue when readers click on links and make purchases, which is why you might see so many gaming websites recommending hardware.

Both traditional advertising and affiliate marketing require traffic, and as social media became less dependable for attracting readers, the importance has transitioned almost completely to Google Search results. This has led to an enormous editorial shift prioritizing what readers are currently searching for: release date news, how to guides, achievement completion lists, and more "service journalism." This traffic desperation has set analytics-driven blinders around games media's coverage, further limiting reporting on new, unknown games in favor of dependable traffic and, thus, dependable revenue in a chaotic time.

That tension makes every facet of the job harder, according to Nicole Carpenter, veteran games journalist and current Senior Reporter at *Polygon*. There are more games coming out every day than she could cover in months, so the realities of the job and the narrowing of games writers mean less coverage.

"As the industry shrinks, it's really hard to balance [covering] the big games that everyone's going to read about that pushes publications forward – that's the stuff people click on," said Carpenter. "Not that people don't click on indie games or stories about games and the climate crisis, but it's a huge balance."

Online media's foundation is periodically destabilized by Google's regular changes to Search. Annual algorithm updates radically alter how media sites are favored, leaving even the biggest sites scrambling to switch up article-writing tactics and broad coverage strategies to recover from plummeting traffic. Many smaller sites don't recover and fold. When smaller sites die, so does coverage for less mainstream niches within gaming: TouchArcade, one of the biggest sites covering mobile games that's been publishing since 2008 shut down toward the end of 2024,[11] citing financial troubles.

The decline of smaller gaming sites means fewer voices, especially those with atypical paths into writing and journalism who didn't have the resources and privilege to enter the media industry through traditional means (expensive graduate degrees, unpaid internships, and low-paid entry-level jobs in high-cost-of-living cities). Mass media layoffs and industry culture hostility, which disproportionately affect women and people from marginalized communities, have driven those voices out of journalism. The media desperately needs those diverse perspectives to appreciate and champion new and novel games that change how we play.

If developers feel that it's harder to get the media to pay attention to non-AAA games, especially those in nascent niches like sustainability, they aren't wrong. Games journalism is smaller, more desperate, more demoralized, and less capable of covering new and exciting games. At the time of this writing, artificial intelligence tools imposed on the games journalism industry by parent companies are poised to wreak havoc on already decimated newsrooms. Again, a dynamic that the game industry knows all too well.

Both the journalism and game industries present a strong facade, a thin veneer of success that benefits only the systematic and corporate interests while abandoning the people who make those institutions possible. But, in the crumbling ruins of journalism and the dwindling career safety of our respective industries, there still persists the determination to keep making and writing about games—because, if you'll excuse the gallows humor, we surviving cockroaches won't let those shareholder-prioritizing bastards take that passion and purpose away from us. Perhaps that leads to a clarity of importance to use gaming as a helpful lens for players to grapple with climate change. All is not lost if we keep doing the work, because people are still receptive to—if not eager for—playing games that reflect their current experience and simulate possible futures. Games are about beating the odds to succeed, and enabling gamers to play out scenarios like climate change can be instructive and potentially empowering. Media outlets can guide players to and through the experiences that will help them engage with the climate-stressed reality they live in. Some eco-themed games have come out and found their audience, with or without the media. Here are examples of successes and how they came to be.

CASE STUDY FOR TERRA NIL: WHERE STRATEGY MEETS SUSTAINABILITY

Terra Nil was just a demo on the independent gaming site Itch.io, created during a Ludum Dare game jam, when Robbie Paterson, a marketing manager at games publisher Devolver, saw it in March 2020. He immediately emailed his coworker asking if she knew about it. Yes, she replied, they were already talking to the game's developers at the South African studio Free Lives.

Coincidence aside, Paterson and others at the publishing company quickly saw that, yes, the game's sustainability themes and mechanics were central to its gameplay, but it was also shaping up to be well-crafted and fun to play.

Terra Nil is a strategy game where players take a square swath of dead land and use technology to restore it to a natural paradise, awash in flora and fauna. At the end of every stage, players scoop up all their ecology-reviving machines and take off, leaving the natural world pristine and unmarked by mankind. Another advantage, then, was that it subverted its own genre as a "reverse-city builder," a moniker attributed by YouTubers and early bloggers writing about the game while it was still a prototype.

That tagline, established before Devolver was involved in marketing the game, also excited the publisher, showing a clarity of mission and a sense of what players would find novel. The publisher recognized that the game promoted the idea of recycling, sustainability, and biodiversity. Paterson speaks to the issue, stating that

> not educating people necessarily, but simplifying what could be very complex issues into a fun video game that people could just enjoy at face value. But it's not hard to see the true intention of what they want to say about the world.

The "reverse-city builder" was such a potent pitch that Devolver felt confident in looking beyond just marketing to games media, also reaching out to academics and science-related publications. Success was mixed, Paterson said; when the outreach started, they found non-games media was more interested in stories with more tangible results—like, would production of this game somehow lead to action in the real world? So, they changed their strategy.

Free Lives' Alfred was a fan of a particular charity that benefited wildlife in South Africa (where the studio is based), so profits from game sales were pledged to that charity. But Devolver saw another benevolent promotional opportunity: Paterson recalled reaching out to influencers and streamers to match hours they streamed playing *Terra Nil* with trees planted. Tangible.

Successful outreach resulted in coverage from publications like *New Scientist, The Guardian*, and other non-games media. And the charity impact resulted in another coverage phenomenon. Normally, games media readers want to hear about upcoming games up to the moment they're publicly launched before turning their attention to the next exciting game, so Devolver's typical strategy has been to intensify focus on this period—but *Terra Nil* had some trailing coverage even after the game launched.

On the whole, Alfred felt positive about partnering with Devolver, not the least of which was freeing him and his team up to work on the game and leave the marketing to their publisher: "Devolver were incredibly useful in helping us realize the best public face to put on the game."

Alfred was also careful to acknowledge how the timing of *Terra Nil*'s launch contributed to its success—namely, that it arrived in the middle of the pandemic, when people were spending more money and time on games. On the other hand, it also launched late enough to touch a nerve with gamers. "I don't think Terra Nil would have been as resonant a decade ago; I feel the world is slowly, belatedly waking up to the climate crisis now, so I'm glad *Terra Nil* is out when it is," Alfred said. It wasn't just Devolver spreading the word, either: back in its prototype stage, the game made around $25,000 in donations on Itch.io. "If I was being cynical, I would say something like people felt like giving us money was helping the environment," Alfred said.

Ultimately, *Terra Nil* was a success for Free Lives and Devolver, and it gave Alfred more perspective on how sustainability games can succeed—namely, by being fun games first.

> Now, I believe quite strongly that you make a fun game and people will want to engage with it and then you can slip in the message, rather than you start with the message and then people are turned off from the beginning,

Alfred said.

Use a game's sustainability themes to attract more attention, divert some revenue to go toward a charity that synergizes with your climate theme—this is the next version of the age-old advice that you should be thinking of how to market your game while still developing it. But it also reflects something else: a story honed over time to explain the success of a game by leaning into its unconventional elements, genre subversion, and creator savviness to explain how it appealed beyond genre. Even the story told to this author is, itself, a crafted version of events playing up uniqueness in content and marketing. The narrative around the game can be just as powerful an appeal as the game itself.

Case Studies: *Outbound* and *Spilled!* on Media Coverage Ahead of the Release of Sustainable Games

Terra Nil came out in 2022 and its makers have had time to reflect on the role journalism played in spreading the word about the game. But how are developers of sustainable games that haven't yet been released preparing for games media coverage? Developers from two of these games—Square Glade Games' Tobias Schnackenberg of *Outbound* and Lente Cuenen of *Spilled!*—offered their perspectives about both the consideration of and indifference to games media attention, respectively.

It's worth noting that these interviews were done in October 2024, many months before the release date for either game. Their perspectives shed light on how developers of games with sustainability themes value (or don't) how games media can help spread the message of their games in the leadup to release.

Schnackenberg's game, *Outbound*, is an open-world adventure allowing players to live out their dream of taking an electric camper van (styled after Volkswagen's famous Transporter microbus) into the outdoors to craft and build their cozy life. To power their dream-home-atop-a-van, players need to harvest local materials, grow food, and build green energy generators harnessing wind and solar to power their off-the-grid lifestyle. The game undeniably incorporates sustainability elements, but this was less a didactic choice to model sustainable practices and more to align with where Schnackenberg and his team saw trends moving. Which isn't to say the team doesn't believe in its message, Schnackenberg assured—just that indie studios need to be shrewd in committing to the right vision to ensure they succeed.

After finalizing a prototype and getting ready to announce *Outbound* to the world in February 2023, Schnackenberg and his team reached out to media to get coverage, expecting few responses. They felt they got lucky when IGN replied with interest in the game and an offer to host the game's reveal trailer, which racked up over 469,000 views as of writing, on the brand's 18 million-subscriber YouTube channel. That led to a domino of coverage from other outlets ("That's usually how it goes – if somebody big posts something, then the smaller ones will follow," Schnackenberg said). The game particularly resonated with journalists who wanted to live out the "van life" fantasy the game was evoking, he added. But the big surprise was among overseas gamers: *Outbound*'s outreach material was translated into ten languages but not Japanese, yet interest in the game blew up in Japan after games site Automaton covered it.[12]

After announcing the game, it became clear that *Outbound* was tuning into the post-pandemic-era yearning to decamp from cities into the wilderness as well as climate anxiety-related desire for sustainable living. Tapping into this zeitgeist grew the game's pre-release fanbase and attracted media attention.

> Outbound is hitting multiple of these relevant topics that people like to talk about at the moment, and so I don't think it's a coincidence that we were able to [get coverage] from multiple news outlets and able to gather quite significant fan base with it, because people are interested in the topics,

Schnackenberg said.

While Schnackenberg and his team figured the game would have several overlapping appeal vectors, they were surprised to find their game appealing to cozy gamers, too. Yet, while he maintained that "the van life fantasy sells better than the sustainability fantasy," the latter is baked into the game's design. Different areas where you park your electric van are powered by varying amounts of solar and wind energy, and there are lots of landmarks that have sustainability topics incorporated into their storyline. This combination is highlighted in the game's marketing, too: the promotional art prominently features its classic-styled van topped by multiple wind turbines and solar panels.

Square Glade Games specifically targeted a mix of journalists and influencers (an outreach ratio of 1:2, respectively) in its preview stage, but once the alpha is released (planned for the middle of 2025), Schnackenberg and his team will switch to mostly reaching out to the latter. "Then we will probably target like 90–95% influencers, because we want them to play the game. We want them to stream the game and actually create content about it. That's not something media would usually cover," Schnackenberg said. They'll return to the mix of journalists and influencers when the game releases in 2026, he said.

Given the small size of Square Glade Games, Schnackenberg wears many hats, including marketing. His best tip is that relationships with contacts, including in media, are the most important thing to have. Those will get better responses than cold emails to a million random addresses, he said. What he and his studio haven't tried yet is reaching out to non-games media, since it's rare for indie games to get coverage from them. But for the *Outbound*'s recent Kickstarter, Schnackenberg found success with local press because it's not common for game creators in the Netherlands to fundraise; the effort raised over $300,000 and became the second most successful Kickstarter campaign in the country's history, he said, garnering local newspaper and radio attention.

By contrast, Lente Cuenen, the solo developer behind the upcoming indie game *Spilled!*, has done very little media outreach. Instead, she's focused on sharing game updates on social media to build a community. "I post on social media and then people seem to write about me. They see my stuff there, and then they think, 'Oh that's a cool story' and they write about me," Cuenen said. She also sometimes works with a public relations firm, Pirate PR, to do outreach on the game's behalf.

Spilled! is a cute game where the player pilots a boat around canals to suck up oil slicks, which can be deposited to upgrade the vessel for better cleanup. As the gunk is removed, the water clears up to show the vibrant plant and animal life below the surface and on the shores. The waterways are lined by houses and wind turbines, which resemble the look and landscape of the Netherlands. While Cuenen didn't set out to make an eco-conscious game at first, her deep care for the environment and nature influenced her while designing and refining *Spilled!*, which she started making at the end of 2022.

Yet, Cuenen's outreach mostly revolved around her community and fundraising for the *Spilled!* Kickstarter. "There's been press writing about it, but it's either through Pirate PR or them seeing my stuff on Twitter. It's not much of my doing," Cuenen said.

> And I like it this way, because you can focus on the game itself and make it the best
> it can be, and then hopefully it speaks for itself and it just spreads on its own right.
> I think that's the best.

Cuenen's brand is her outreach, growing her Twitter (now known as X) account to over 22,000 followers, the game's Discord channel with around 2,000 members, and sending a personal newsletter to around 2,000 subscribers filled with game updates and personal details about her experience living on a boat. Recently, Cuenen noted that events she has participated in—like the Zero Carbon Quest event on Steam at the end of 2024—bring in more wishlisting players than any Twitter post ever did.

Most developers of indie games planning to release on PC heavily prioritize getting prospective players to wishlist their games on Steam, a low percentage of which translates to sales, thus serving as a predictive metric for launch success. Developers plan outreach accordingly to catch player attention ahead of launch, including reaching out to media—though the latter wasn't part of Cuenen's plan either. "I haven't really found a huge correlation between coverage and wishlisting," she said. "It could be nice to get people to know about me and my brand, my studio if you will to get people following along. That would be nice, but it's not a big focus for me."

While Cuenen is grateful for the help that the Pirate PR firm has done to help market the game, she feels it's a nice bonus rather than a necessity, which she admits may not be a common way for indie developers to think. "I don't know, maybe I'm just really lucky that the game has been kind of marketing itself," she said. "Opportunities just come my way, and maybe I'm really lucky, but I haven't felt as much of a need to get a lot of help with marketing or hiring a company for this."

Lucky as Cuenen admits she may be, especially since *Spilled!* is her first major game release, she doesn't plan to change her approach for the next game, and will continue to prioritize building community over fielding media attention.

As a last point of trivia, both of these developers who chose to make games with sustainability themes are from or currently live in the Netherlands. Given they were inspired in varying ways by the environment, culture, and lifestyle of Dutch living, this does not seem coincidental. Schnackenberg noted that he was inspired by the plentiful wind farms dotting the rural areas he'd drive through around where he lives in Groningen, a city in the northeast Netherlands. Through her Twitter posts and newsletter blasts, Cuenen shares her boat lifestyle plying the waters of her home country, delighting in living—and making games—in nature.

PERSPECTIVE: ALEX AVARD AND THE FIRST SUSTAINABILITY GAMING NEWSLETTER

Games journalists follow reader trends—the more ahead of the curve, the better. Though I had discovered the growing world of eco-gaming in 2023, a wiser journalist, Alex Avard, had been covering the intersection of gaming and sustainability since 2018, when he read that year's influential *Intergovernmental Panel on Climate Change report*[13] on the 1.5°C temperature increase. As a writer for the UK-based publication GamesRadar, Avard told me that he had pursued these stories between his normal duties covering new conventional games; as a testament to that publication, his editors never dissuaded him from covering sustainable gaming. Because GamesRadar was a "celebratory site" that leaned toward more positive coverage, Avard shaped his writing accordingly.

"I'm of the persuasion now that actually there's enough doom and gloom out there in the kind of general narrative towards the climate crisis," Avard said. "What people need are stories not of false hope, but stories that can inspire people to take action with the kind of psychological resilience they need so that they're not feeling that it's pointless because everything's going down."

For Avard, writing these sustainability gaming stories was a way of taking action himself while also providing a lens for gamers to engage with the climate crisis. Avard states-

> I wanted to write the kind of stories that I would want to read about to make me feel a little bit less existential about the entire situation and feel encouraged that my favorite hobby and pastime did have a part to play in this overall fight against the climate crisis.

Avard follows up by claiming, "That there were ways I could get involved as well by supporting games that were doing stuff on this agenda."

Avard continued writing gaming sustainability stories until he left GamesRadar in 2021 to take a position at a non-profit, but still felt passionate about sustainability gaming, which led him to start a newsletter on the subject: Play Anthropocene.[14] Every two weeks, the newsletter rounded up any new releases, reports, or other news related to sustainability gaming.

Avard noted that, at times, it was difficult to find enough to write about, and a fortnight could go by with barely two newsletter-worthy items to include. But the precedent it set was crucially important, spotlighting efforts and games with the occasional interview that provided game developers and experts space to go deep on the complexity of sustainability. It was also a dedicated pulpit Avard used to critique the game industry's slow progress toward taking responsibility for its impact on the climate.

Growth was steady over the next two years as readers continued signing up for the newsletter. Then Avard accepted a new job at Playing for the Planet, the UN-aligned non-profit that coordinates green initiatives in the game industry and encourages developers to adopt more sustainable practices. But given the role's work engaging with studios and publishers, Avard felt it wouldn't be appropriate to continue the newsletter. It's been on indefinite hiatus since 2023.

While Avard doesn't rule out restarting the newsletter, its impact is still important as an editorial space of *consistent coverage*. Avard's persistent writing at GamesRadar was the exception to the rule that sustainable gaming stories remain extremely rare and aren't regular enough for a staff or freelance journalist to roll up into a beat. Nevertheless, the newsletter collected relevant reports and updates on eco-themed games to serve as a timeline record for the efforts in that space, and allowed him to follow up on older news to show progress on issues.

Too soon do we forget eco-pledges from big companies that become forgotten, and there's no beat reporter holding the biggest publishers in the industry—and thus, the biggest polluters—to their promises. This is especially worrisome considering the onset of AI, where we've seen tech giants' declining energy emissions reverse course as massive compute needs cause energy use to skyrocket. Advocates for sustainability in the game industry hadn't even extracted lower emissions commitments from big publishers, studios, and console creators (aside from tech companies-cum-game publishers Microsoft and Sony); now there's no pre-AI pledge to hold these potentially big emitters too.

But there is also a scarcity spiral at work: without many climate-related games, there's less justification for a beat reporter covering the niche. More games speaking truth to the moment of climate stress can and will catch the attention of journalists and, working with them, carve out a space for these experiences to reach their audience. And given the thematic resonance between these games and sustainability itself, it's easy to see how such reporter coverage would extend to scrutinizing the game industry as a whole. The benefits extend beyond mutual support—they can help us make the game industry answerable to the needs of the planet.

PERSPECTIVE: LEWIS GORDON, THE MOST PROLIFIC SUSTAINABILITY GAMES REPORTER

In 2017, UK-based games reporter Lewis Gordon started writing about video games and noticed some were unmistakably incorporating radically changing climate in their narratives. That year saw the indie game The Flame in the Flood set in a world waterlogged by apocalyptic floods, as well as AAA blockbuster Horizon Zero Dawn set in a robot-dominated land regrowing after global calamity—and from there, Gordon fell into writing about all the ways game development and hardware manufacturing intersect with the climate crisis.

Gordon has written a number of stories on the subject for a variety of publications both in games media and mainstream publications. From breaking down the environmental impact of the PS4 console for tech site The Verge[15] to a feature about sustainability-themed games in *The Guardian*[16] to an overview of extreme weather as a game mechanic for *The New York Times*,[17] Gordon has successfully brought the intersection of climate change and gaming to mainstream media.

Writing for papers of record like *The New York Times* and *The Guardian* is an accomplishment in itself given their prestige, because mainstream publications don't regularly publish articles on video games (The Guardian has one video game editor on staff, *The New York Times* has none). Gordon noted that he doesn't write for more enthusiast gaming sites, and sticks to the publications that feature well-reported stories, which have included big tech and mainstream sites. From his perspective, this has insulated him somewhat from the closures of more niche enthusiast gaming sites, though he would still like to see gaming-focused outlets cover sustainability gaming topics, even if that's trickier with their audiences who come for games updates.

By sticking to more general interest publications, Gordon hasn't found it too difficult to get his story pitches in this intersection of climate and gaming approved—possibly because he's one of the only ones doing it. "Editors out there do seem incredibly receptive to this stuff. I think maybe I find it easier because not many people have been [writing in this niche]," Gordon said.

Yet, lately Gordon has been hitting a wall with his coverage of sustainability gaming. "There are only so many kinds of angles that I can necessarily offer on the subject before it feels like I'm repeating myself," Gordon said, noting he doesn't know how many times he can keep saying that hardware or game development negatively impacts the planet in terms of emissions. He's still looking toward new games coming out with sustainability themes, but the bar is higher now, especially now that there are initiatives followed by studios seeking to get their operations closer to net-zero emissions.

Games that are simpler "environmental fables" about cleaning up vague pollutants have become common enough not to warrant coverage on their own, Gordon said. When Abzu came out in 2016, it was novel and featured oceans teeming with life and felt like an environmental allegory, he noted, but years later audiences have seen climate change in real time. "If we cloak a lot of this stuff in more generalized visual abstracted from our own world, it can lose some of that kind of potency," Gordon said. The 2023 AAA game Avatar: Frontiers of Pandora, by comparison, features players taking on a pseudo eco-terrorist role destroying human infrastructure and seeing natural life return in real time: "I was actually quite surprised by how much bite that had," he said—noting that the game came out a year after the 2022 film How To Blow Up A Pipeline. The appetites of the environment-sympathetic public have shifted.

But the core of any game has to be appealing, both in mechanics and gameplay, Gordon reiterated.

THE BEST WAY TO PITCH, FROM JOURNALISTS

While the current shrunken state of games media leaves little room for anything beyond news and reviews, for better or worse, climate change is saturating the mainstream and becoming unignorable. As sustainability grows in the public awareness regarding electric vehicles, green energy, clothing, supply chains, weather, and broader environmental concerns, it's easier for journalists to draw parallels with their pitches.

"I think that readers want to see more in-depth writing about how the video game industry impacts the climate crisis," Carpenter said. But she cautioned that the issue could be touchy for readers who are likely uncomfortable interrogating the role of gaming in the climate crisis. Still, it can be challenging in a good way, but only for folks who have a concrete idea.

It's possible—and as games media wakes up to sustainability gaming, very probable—that developers will interact with journalists who are not informed of all the work already done in the space. They will come in a variety of attitudes, some brazenly ignorant and others humbly aware of their lack of knowledge. On a personal note, having been in this situation in 2023, I beg your patience with the latter.

As with other subjects that have been "discovered" by journalists unaware of the work happening years before, there are ways to frame sustainability gaming, and your project specifically, in ways that will make their importance and impact clear. Broadly speaking, from gaming media's perspective, there are two angles to approach when writing about gaming sustainability: how gaming is handling emissions and how it's handling integrating sustainability themes.

Emissions are a more complex topic and tricky to convey in impact, so much so that Avard wouldn't advise developers to focus on decarbonization in their outreach. In my own 2023 article[18] on the gaming industry, I can attest to the lengths it took to convey the complicated scope and scale of the industry's emissions footprint. I extensively benefited from Australian analyst Ben Abraham's reports and book *Digital Games After Climate Change*.[19] His work has collated the most complete picture of how much publishers, studios, and hardware sellers are contributing to the game industry's overall emissions impact. Given the piecemeal data provided by the industry—which is not standardized in estimation and largely voluntary to disclose—his reports are supplemented by projection, resulting in ranges for the varying scope 1, 2, and 3 emissions. Abraham's work is discussed more thoroughly in Chapter 12 of this book.

What Abraham delivers is clarity and relativity, both precious currencies to journalists: in chapters of his book and his newsletter, *Greening the Games Industry*,[20] Abraham provides methodology for game makers to convey their climate impact compared to the rest of the industry. For instance, Abraham's book highlights UK studio Space Ape Games and Finnish studio Rovio as examples of studios proactively sharing emissions data compared to contemporaries and much larger publishers and console makers. Given the industry's reluctance to share their climate impact, sharing concrete figures accounting for all the ways a company affects the planet is the transparency journalists want to see.

Journalists may laud cutting-edge graphics, but it's far from essential for a hit game—just look at *Stardew Valley*,[21] *Celeste*,[22] and *Balatro*,[23] which are beloved for strong gameplay and unique art styles. Shrewd studios can remind journalists that development decisions to make less graphically intensive games result in less computation to play them. Reducing specs requirements through optimization literally lowers emissions, as players using less advanced computers with older graphics cards or consoles aren't emitting as much as those running the latest Nvidia and AMD GPUs. The Nintendo Switch, which is essentially a tablet, is even more efficient than its console competitors.

> I think of games like The Legend of Zelda: Breath of the Wild or Tears of the Kingdom which don't go for high-end realism but receive high amounts of critical acclaim and sold millions of copies can run on the Nintendo Switch, and compared to the PS5 and Xbox Series X has far less of an energy drain,

Avard said.

The games platform that uses the least energy is phones, and mobile gaming is, roughly speaking, the most efficient way to play. Mobile game developers can appeal to journalists in their lower contributions to climate change.

The other avenue of sustainable games coverage is titles with eco and sustainability themes woven into their gameplay, narratives, and/or vibes. This is a broad umbrella, and could cover vaguely associated games, such as those set in a post-apocalypse after a climate catastrophe that don't otherwise engage with climate. But games that thoughtfully engage with and integrate sustainability into their games should lean into the topicality of their subjects and novel mechanics. Like any art form, games are a lens to grapple with real-world issues, and developers can appeal to journalists' desire to cover games with resonant issues.

We're far past the days of *Math Blaster* and baldly educational games, but there is intriguing value in conveying the spirit of the climate struggle even if it doesn't reflect the factuality of climate repair or rescue. To wit, *Terra Nil*'s technology is futuristic and fantastical, as our current weather manipulation is dubiously effectual, but the process of procedurally restoring layers of ecosystem is itself a helpful exercise to temper player expectations.

As opposed to the negative story of gaming's footprint, it's easier for developers to sell the so-called handprint of gaming's potential for positivity, Avard said: "The potential of games as activators for the climate movement is really a much sexier story to me, and there's loads to cover in there."

Of course, gameplay appeal also matters—developers should lean into the unique mechanics and details of their game in seeking coverage, journalist Carpenter advised. This could take the form of publishing dev blogs or making developers available to speak to journalists. She recalled a story she wrote for *Polygon*[24] about Nintendo's 2023 Tears of the Kingdom, in which developers not involved in the game gave insight into its design—and readers ate it up. "That kind of coverage is really interesting to me, and I would be interested to see how it could be applied to games that have sustainability themes or development," Carpenter said.

In the sea of emails a games reporter receives, an individual developer's story can pique more interest, Gordon agreed. Interesting and weird ideas baked into a game's central conceit, along with strong visuals to tell your game's story in an image or two, can grab a reporter's attention. Gordon pointed to the upcoming Wood & Weather by Australian studio Paper House, which wraps its environmentally conscious weather-controlling gameplay in a delightful wooden toy aesthetic, as an example of a game that ticks these boxes for him.

Obvious as it might be, online visibility matters in gaining attention. In addition to actively reaching out to journalists, developers should consider posting videos on TikTok and YouTube or streaming on Twitch as passive outreach that could pique games writer interest, Carpenter said: journalists are paying more attention to these video-focused discovery spaces that attract fans who make interesting, new games go viral.

Most importantly, journalists have to be able to reach developers—who could be missing out on helpful coverage if they don't make themselves easy to find online. "One big thing that people forget is to make it easy to contact you," Carpenter said. Be visible on search and social platforms, have your email address very visible in profiles, and don't be afraid to continue following up with journalists who haven't responded to inquiries. "That's never going to be annoying to me, because there are simply too many emails to respond to. It's not annoying at all – it's important," Carpenter said.

In any niche of journalism, it's crucial to research writers and appeal to the ones who cover similar things. For developers, this means looking for writers who have written about games like yours—but it's on you to explain why yours is different.

Pragmatic Points/Summaries from Case Studies

- *Terra Nil*: Align eco-gameplay with thematically resonant actions in the real world with a marketing angle, like planting trees for every hour streamers play the game, to appeal beyond games media.
- *Outbound*: Cold email big games media sites, which could multiply your exposure. Also, don't ignore foreign audiences and journalists, who could be more receptive to your game.
- *Spilled!*: Focus on growing your community and tell your personal story on social media as it aligns with your game's eco-themes. Journalists can and will organically notice.
- Working with PR companies can free up indie developers to focus on their game, and they can spitball novel marketing strategies to draw media attention.
- It can help to strategically send codes to different ratios of journalists and streamers at different times: journalists when you need critical attention in early stages, streamers when you need players around launch.
- Tap into gamer desire zeitgeists: escape, cleaning, leaving no trace to live harmoniously with nature.

ADVICE FOR JOURNALISTS

Just as developers get savvier at speaking to journalists with more appealing angles and hooks, so does games media need to adapt its reporting and coverage to understand the value of sustainability gaming. Journalists are, by nature, wary of a company's positivity out of concern that self-effacing progress masks deeper dysfunction, but there's more to covering sustainability gaming than assuming it's all greenwashing.

Some of this adaptation is resisting the urge to follow the presumed appetites of gamers for better graphics at all costs, an expectation that has arguably contributed to sky-high expectations for new AAA games that have extended development time and costs.

"I think there's something to be said, and this is certainly a wider conversation around economics and sociology around decoupling the climate movement from growth," Avard said.

> I like the idea of a new movement of games that can do more with less and not push for that energy-intensive operational power and just provide amazing games that don't need to draw massive amounts of energy from the grid.

And as a corollary to Gordon's desire to learn how game developers' personal stories led them to unique game designs, it's also up to journalists to discover the context behind creation. Be curious: every developer has a story, and often the most interesting ideas come from where they're located. When Gordon interviewed the Free Lives developers behind *Terra Nil*, their stories about running out of water due to South Africa's draughts gave his story so much more flavor. These can provide story hooks that editors will be more attracted to than simply covering a new game.

Games media, too, bears some responsibility for decades of pushing for the highest performance in games and fallaciously equating that with quality. Seasoned journalists are soberly aware of this impact, but novices continue to echo the seemingly majority gaming opinion that new games must meet a baseline of graphical sharpness and sustained high framerates. I implore my fellow journalists to take this to heart; the "I want shorter games with worse graphics and I'm not kidding" meme was originally written by one of our own, Jordan Mallory, in a 2020 post on X (formerly Twitter).[25]

Due to the unfortunate waning of employed games journalists, there's a regrettable lack of enthusiasm and funding for reporting on the industry itself. But as climate change intensifies its impact on the world, game development—especially in regions hit harder by storms, droughts, and other effects of a warming planet—will suffer. How the industry is contributing to these problems is just as important as quarterly financial results.

> I do think there is a responsibility here, not just in games but generally in the press, in keeping people accountable to this stuff on how [companies] are doing and where they need to go next in terms of decarbonization and plastics and packaging and all of this kind of stuff,

Avard said.

Based on the journalistic work done by Avard, Gordon, and others, it's clear that reporting on sustainability gaming is possible in mainstream publications or in more focused newsletters. The limiting factor is the lack of publishers willing to see the value of covering the intersection of gaming and the climate doom of our age. Those who have pushed to cover this overlap do so alone, eventually burning out from the strains of the media industry and lack of support, even leaving entirely. Yet, even in reaching out to them for this chapter, they were not bitter about their coverage, insisted upon its necessity, and believe readers are deprived without it.

Though fewer in number, there are reporters who still dig into the game industry—they just primarily cover its overall health, focusing on layoffs and studio closures. They have yet to turn their energies to the service journalism of sustainability issues or illuminating the energy consumption of making and playing games, of responsibly turning a mirror on the entertainment industry and its enjoyers. It's a failure not without sympathy, as these journalists prove their value to publishers with scoops and exclusive stories rewarded by reader traffic. But surviving in a media industry on life support leads to a reactive news sense based on desperation, not aspiration—appealing to what's come before, not what could be.

Finally, perhaps the best of these reporters are somewhat hindered by the ideological bulwark that separates journalism from the game industry. The most esteemed and principled journalists rightly point out that they are not and should not be considered a part of the game industry, a perspective that preserves objectivity for the press to present their truth to readers without undue influence. This separation is more important than ever, as a resurgence of the noxious and pathetic Gamergate movement again targets women and media writers from marginalized backgrounds to challenge their competence and integrity with baseless, absurd accusations. There's concern over showing preference or favor; hence, remaining at a distance from the industry, telling stories traditionally popular with readers.

But, between journalism and the game industry, there remains an existing partnership to convey gaming to readers. This mutually beneficial relationship can be harnessed to tell a relatively new category of story that Avard, Gordon, and others have pioneered—engaging with climate change through gaming. What does it mean to interactively grapple with a world more challenging by the day? Journalists do not have to be subject to the game industry's agenda, nor compromise their professional integrity by sparing game makers from deserved criticism, but they can believe in a mutual goal of spreading the word about how the medium of gaming can, and may be, rising to the moment. Will they, too?

NOTES

1 Lumb, David. "Video Games Are Finally Waking up to Climate Change." CNET, December 30, 2023. https://www.cnet.com/tech/gaming/features/video-games-are-finally-waking-up-to-climate-change/.
2 Free Lives. Terra Nil. Devolver Digital. 2023.
3 Lente. Spilled!. Lente. 2025.
4 Square Glade Games. Outbound. Square Glade Games. In Press.
5 Lumb, David. "Video Games Are Finally Waking up to Climate Change."
6 Zwiezen, Zack. "Tencent Guts Its Gaming News Site, Firing Staff without Warning." Kotaku, September 15, 2022. https://kotaku.com/fanbyte-website-layoffs-tencent-staff-g4tv-future-1849540893.
7 Dealessandri, Marie. "Vice Shutting down Waypoint." GamesIndustry.biz, April 28, 2023. https://www.gamesindustry.biz/vice-shutting-down-waypoint.
8 Hume, Mike. "What Launcher Meant, and Means, for Games Journalism." *The Washington Post*, March 31, 2023. https://www.washingtonpost.com/video-games/2023/03/31/launcher-mission-end/.
9 Plunkett, Luk. "Goodbye Kotaku Australia, You Were a Very Good Website." *Aftermath*, July 8, 2024. https://aftermath.site/kotaku-australia-closed-layoffs.
10 Peters, Jay. "Game Informer Is Shutting Down." *The Verge*, August 2, 2024. https://www.theverge.com/2024/8/2/24212016/game-informer-shutting-down-layoffs-gamestop.
11 Nelson, Jared. "TouchArcade Is Shutting Down." *TouchArcade*, October 11, 2024. https://toucharcade.com/2024/09/24/toucharcade-is-shutting-down/.
12 Fujii, Yusuke. "キャンピングカー生活ゲーム『outbound』Steam向けに発表。オープンワールドを冒険サバイバルしつつ愛車をカスタム、気ままな自給自足生活." AUTOMATON, February 13, 2024. https://automaton-media.com/articles/newsjp/20240213-282309/.
13 IPCC, 2018: Summary for Policymakers. In: Global Warming of 1.5°C. An IPCC Special Report on the impacts of global warming of 1.5°C above pre-industrial levels and related global greenhouse gas emission pathways, in the context of strengthening the global response to the threat of climate change, sustainable development, and efforts to eradicate

poverty [Masson-Delmotte, V., P. Zhai, H.-O. Pörtner, D. Roberts, J. Skea, P.R. Shukla, A. Pirani, W. Moufouma-Okia, C. Péan, R. Pidcock, S. Connors, J.B.R. Matthews, Y. Chen, X. Zhou, M.I. Gomis, E. Lonnoy, T. Maycock, M. Tignor, and T. Waterfield (eds.)]. Cambridge University Press, Cambridge and New York, pp. 3–24. https://doi.org/10.1017/9781009157940.001.

14 Avard, Alex. "Play Anthropocene: Alex Avard." *Substack*. Accessed April 25, 2025. https://playanthropocene.substack.com/.

15 Gordon, Lewis. "The Environmental Impact of a PlayStation 4." *The Verge*, December 5, 2019. https://www.theverge.com/2019/12/5/20985330/ps4-sony-playstation-environmental-impact-carbon-footprint-manufacturing-25-anniversary.

16 Gordon, Lewis. "Can Video Games Change People's Minds about the Climate Crisis?" *The Guardian*, January 26, 2023. https://www.theguardian.com/games/2023/jan/26/can-video-games-change-peoples-minds-about-the-climate-crisis.

17 Gordon, Lewis. "Extreme Weather Gives Video Games a Powerful Adversary - The New York Times." *The New York Times*, October 4, 2024. https://www.nytimes.com/2024/10/04/arts/extreme-weather-storms-video-games.html.

18 Lumb, David. "Video Games Are Finally Waking up to Climate Change."

19 Abraham, Benjamin. *Digital Games after climate change*. Cham, Switzerland: Springer, 2022.

20 Abraham, Ben. "Greening the Games Industry." *Greening the Games Industry*. Accessed April 25, 2025. https://gtg.benabraham.net/.

21 ConcernedApe. Stardew Valley. ConcernedApe. 2016.

22 Maddy Makes Games Inc., EXOK Games. Celeste. Maddy Makes Games Inc., 2018.

23 LocalThunk. Balatro. Playstack. 2024.

24 Carpenter, Nicole. "Why Tears of the Kingdom's Bridge Physics Have Game Developers Wowed." *Polygon*, May 25, 2023. https://www.polygon.com/legend-zelda-tears-kingdom/23737921/tears-of-the-kingdom-bridge-physics-game-devs-explain.

25 Mallory, Jordan @jordan_mallory. "i want shorter games with worse graphics made by people who are paid more to work less and i'm not kidding." Twitter, June, 29 2020, 02:07. https://x.com/Jordan_Mallory/status/1277483756245442566?lang=en.

Building a Green Game Movement

11

An Overview of Pro-Environmental Efforts in the Game Industry and Beyond

Valley Lopez

ABSTRACT

This chapter establishes a framework for understanding environmental advocacy efforts in the video game industry by examining the emerging "GGM." Drawing from interviews with 15 members of the International Game Developers Association Climate SIG and its affiliates, the research illuminates the strategies, action areas, and challenges facing the GGM. Recommendations for strengthening pro-environmental efforts in gaming are offered throughout.

KEY TAKEAWAYS

- **The green game movement is real—but fragmented**. A growing coalition is pushing for environmental change through games, but lacks a shared strategy or unified vision.
- **Community is the engine.** Collaboration, knowledge sharing, and paid roles are essential to making sustainable game advocacy effective and lasting.

DOI: 10.1201/9781003512400-13

- **Advocacy needs infrastructure.** Passion alone won't sustain the movement—resources, recognition, and coordination are critical.
- **The movement is evolving—fast.** The Green Games Movement is dynamic and expanding, but its long-term power depends on clearer direction and collective alignment.

INTRODUCTION

During the 2010s, a diverse coalition formed to leverage video games as tools for pro-environmental change. Gamers, scholars, activists, policymakers, game developers, studios, and more have joined this effort, which continues to grow and evolve. I refer to this conglomeration as the green game movement (GGM). Specifically, the GGM describes the loose assemblage of people, organizations, and advocacy efforts at the intersection of video games and the environment that seeks to expand pro-environmental knowledge, attitudes, and behavior. Like the broader environmental movement, the GGM is constantly shifting and lacks clearly defined boundaries.[1]

As a researcher at the University of Oxford, I set out to analyze the current state of the GGM from the perspective of developers and game industry environmental advocates. These advocates include those who interface with game industry environmentalism directly but do not create games themselves, such as academics and publishers. I was interested in unpacking two things: (a) how members of the GGM contribute to pro-environmental efforts and (b) the key challenges they are facing. In doing this, I aimed to contribute practical knowledge and insights to empower game industry environmental advocates in their future work.

Along with a review of the academic literature, my research was based on 15 interviews with affiliates of the International Game Developers Association Climate Special Interest Group, or "SIG." The SIG was founded in 2020 as a volunteer interdisciplinary network to foster a "community-based participatory approach" to environmental advocacy among game developers.[2] The central hub of the SIG is an open-access Discord server, where over 1,200 members (and growing) assemble from around the world to share information, resources, and strategies related to climate and gaming.[3] The SIG has a few strict guidelines and no clearly defined leadership structure, primarily functioning as a grassroots forum.

No other space brings together such a wide range of perspectives on gaming and the environment, which makes the SIG a unique portal to consider the GGM. The SIG also contributes crucial knowledge to various sustainability initiatives and plays an essential role in bringing more people into green gaming. At the same time, the SIG is a victim of several of the same challenges that hold back the wider GGM. Addressing these challenges will be critical if the efforts outlined in this book are to be effective.

In the paragraphs below, I utilize my research findings to outline a framework for the GGM. I also suggest ways for those involved in the GGM, particularly game developers, to heighten the impact of their advocacy efforts. These findings are meant to be supported by more specific and in-depth investigations from the other authors of this book. Importantly, I remain wary of proposing any firm solutions, as complex environmental issues will not be "solved." Instead, I hope the insights in this chapter provide a big-picture approach to building the future of the GGM.

SIG Strategies

The SIG's impact stems from three key strategies. The first involves **creating resources and sharing knowledge**. The SIG-developed *Environmental Game Design Playbook* (EGDP)[4] is a perfect example. The EGDP is a plain-spoken, 75-page report that unpacks the theories behind effective player impact and offers overviews of tactics for creating more effective green games. Game Director Andrew Brennwald attested to the usefulness of the EGDP, describing it as "a blueprint of how you make green games." Other SIG-developed resources include the developer-oriented website greengamedesign.com and the *Doing Our Bit* podcast, which focuses on environmental advocacy within the game industry. The SIG Discord channel is also a vital resource, as it offers an open forum for members to share ideas and information.

The second strategy focuses on **community building**, which points to the SIG's role in creating a positive and inclusive space for collaboration. The group unites people from various disciplines around common goals and interests, an important factor in breaking down boundaries between the game industry and other fields. Game Designer and SIG Co-founder Hugo Bille described the group as a forum for "making sure that people are all connected to each other, and they have the means to aid each other in whatever way they can." Megan Carriker (Game Publisher and Head of Marketing) highlighted the welcoming nature of the SIG, noting that it's a "safe space" for advocates who may feel lost or isolated.

The SIG has also played an essential role in **growing the GGM**. Although its membership numbers are impressive, the group's reach extends well beyond the Discord channel. The SIG hosts the highest rated workshop at the Game Developers Conference, and each year hundreds of people attend talks by SIG members. One of the SIG's Discord groups focuses on establishing environmental advocacy groups at game studios throughout the industry, which several members have succeeded in doing. Game Production and Leadership Consultant Grant Shonkwiler noted that these movement-building actions are powerful because they inspire people to take action once they discover a like-minded community.

Although each SIG strategy has value individually, they do not operate independently of one another. Instead, all three occur simultaneously and build on each other in an iterative manner (Figure 11.1). This symbiotic relationship highlights several pathways for engagement within the wider GGM

- Cultivating knowledge-sharing mechanisms within gaming
- Developing inclusive community structures that embrace multiple perspectives
- Expanding the movement through outreach efforts.

While these are not the only possible pathways for strengthening environmental advocacy, they serve as solid strategies for building impact and engagement.

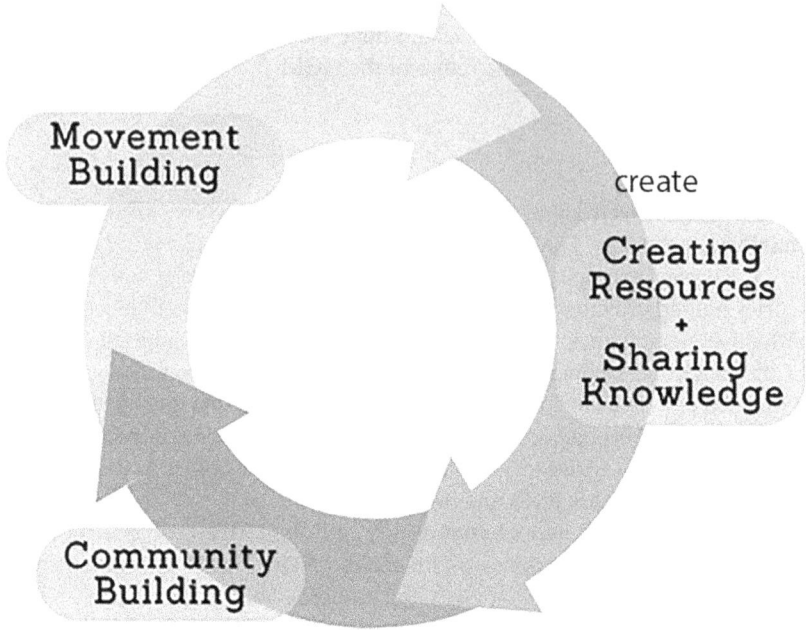

FIGURE 11.1   SIG strategies of environmental advocacy.

(Source: Author).

Pillars of Action

The SIG's strategies provide a foundation for *what* we can do, but *where* should our advocacy efforts be focused? Analyzing the SIG reveals how pro-environmental initiatives in the game industry are organized around four "pillars" of action: **game design, activism, footprint reductions**, and the **metagame**. While these pillars highlight an organizational framework, they also indicate how efforts might be enhanced or reevaluated.

Game Design

Using game design to influence players is the most prominent form of pro-environmental action in the GGM. It was also the most frequently discussed form of environmental advocacy in my research. Below are several key design-related themes that emerged from the interviews.

- **Behavior change**
  - Green game design should aim to change players' behavior rather than only inform or educate them. There is no set method for achieving behavior change, but it is worthwhile to continue seeking the right "formula" for impactful design.

- **Environmental representations**
  - Developers should consider avoiding harmful environmental tropes in games, such as extractivism, post-apocalyptic ruin, and ecological destruction.[5] Portraying more positive visions of the future will likely enhance a game's pro-environmental impact.

- **Beyond climate change**
  - Games should engage with the complexity of environmental issues rather than focusing solely on climate change. For example, game narratives could probe deeper into the structural (i.e., social and political) causes of environmental change.[6,7]

- **Commercial vs. serious games**
  - Many participants questioned whether "serious" green games are impactful if they only reach a small, highly specific audience. This is also a common theme in the academic literature.[8,9] None of the interviewees expressed that producing more serious green games would be better for player change. Instead, they advocated for creating green games with broad commercial appeal.

- **No perfect green game**
  - Andrew Brennwald notes that a single game can't "own the entire behavior change cycle and do it effectively." In other words, no green game can influence all players or change the world. Instead, green games are a powerful *combined* force.

Chapters 8, 16, 17, 18, and 19 discuss green game design strategies in more detail.

Developer Activism

The second pillar of action addresses how we might use games as a springboard to pursue various forms of activism. Game Director Jennifer Estaris has been a key part of this movement (see Chapter 23). She helped establish the "Game Dev Rebellion" channel in the SIG Discord server, which she described as a meeting place for advocates to "try to figure out" how to inspire activism among game developers. Game Researcher and Designer Joost Vervoort also played a crucial role in forming the channel, which he views as a way to "[bring] game developers and activists closer together" so they can use their platform to strategize collectively.

However, it's unclear exactly what role activism plays in the SIG. Hugo Bille noted that the group's intentions are well-placed and widely supported, but he stated, "I'm not sure that we [have] necessarily found the thing that clicks quite yet." Vervoort described the "slowness" of activism as a challenge, noting that Game Dev Rebellion has been a "start-stop" movement because people lead busy lives that prevent them from fully committing to the cause. Studio Director and Game Designer Trevin York pointed out that the instability of the game industry could be preventing more people from getting on board.

Activism is a cornerstone of the broader environmental movement, so what is needed to open it up for those in the game industry? Perhaps it could be useful to expand its definition. Software Engineer and Programmer Nick Schrag describes the range of activities activism encompasses

> The throwing tomato soup at paintings is sort of the bigger risky civil disobedience kind of activism. And then there's the more working within the system, talking to your elected officials, and that whole spectrum.

In other words, you don't have to be out on the street risking it all to be an effective activist. Contacting policymakers, signing petitions, boycotting unsustainable businesses, or supporting other activists are all ways to contribute. There is also room for designing games that depict the power of activism, such as *Alba: A Wildlife Adventure*[10] and *Disco Elysium*.[11] Chapter 21 contains useful insights on this topic.

In general, there is a need for more clarity around activism's place in the GGM. A prominent theme throughout the interviews was that current activism efforts lack structure. This is understandable—there is no guidebook or clear path for industry workers who want to engage in effective activism. To address this, we should work to bridge the gap between game developers and established environmental activists by strategizing how games can contribute[12,13] (see Chapter 6). Ideally, we should develop a framework for activism that ties in with the other pillars of action.

Mitigating the Footprint of Games

The third pillar of action addresses the environmental impact of the game industry. Efforts in this sphere address emissions reduction, resource usage, and the ecological footprint of gaming devices.[14,15] For example, the SIG's "industry reporting" channel allows members to account for the carbon produced by their studio and brainstorm strategies to reduce energy usage. Nick Schrag acknowledges that this is a "less exciting" form of environmental advocacy but argues the industry should put its money where its mouth is and not just focus on player-based initiatives.

Developers might also explore ways to make their games more efficient. Grant Shonkwiler points to the energy requirements of games with high graphical fidelity, stating that "we don't need to be pushing graphics cards to the limit all the time." Game Consultant and Designer Shayne Hayes notes how *Fortnite*[16] was designed to lower the resolution and framerate of a game when it sits dormant on a pause screen, resulting in significant energy savings.

Minimizing the production and consumption of physical hardware is a further consideration, particularly as we navigate an "upgrade culture" that continually pushes us to purchase the newest and most powerful devices.[17] Hugo Bille asks, "How do we run on as old hardware as possible so that we can stop perpetuating this arms race of buying new stuff all the time?" Given that the economic model of gaming relies on continued consumption,[18] there is no straightforward answer to this question. However, Arnaud Fayolle suggests it would help to keep releasing games on all platforms to disincentivize the need to upgrade. Benjamin Abraham's book *Digital Games After Climate Change*[19] has been a vital resource for these sorts of aspirations, and he has also contributed to Chapter 12 in this book.

Efforts to reduce the footprint of games are relatively well-defined, even though these discussions are still in their early stages. To contribute to efforts in this pillar, we should continue sharing information about how players can minimize their impact, how studios can implement sustainable practices, and examine the broader material ecosystems behind digital games (Patrick Prax shares insights on this topic in Chapters 2 and 5). Importantly, we should be mindful that playing and creating games (green or not) is currently at odds with our pro-environmental goals if these activities are underpinned by unsustainable systems.[20]

The Metagame

The final pillar of action concerns the spaces, ideas, and knowledge that exist outside of games themselves—also known as the "metagame."[21] Put simply, the metagame is "the game beyond the game."[22] This somewhat slippery concept includes (but is not limited to) social communities, game websites, social media forums, game strategy guides, merchandise, and the physical spaces where people play games.[23]

The SIG-developed Environmental Game Design Playbook describes the metagame as "a tool for reinforcing an environmental message."[24] According to Hugo Bille, harnessing the metagame creates the possibility of "player attitudes changing from engaging with the community around the game." Trevin York notes how we might "[build] a bridge between the in-game communities and local communities" by directing players to activities such as farmers' markets and foraging. Online forums for green game players could be utilized as well.

Twitch streaming is a particularly promising method for enhancing player engagement, given that the platform attracts over 105 million monthly visitors.[25,26] Relatedly, games researcher Gaia Amadori described how high-profile Twitch streamers played *Minecraft*[27] to raise awareness about deforestation in Poland.[28] As part of a coordinated campaign with Greenpeace, the streamers helped gather 170,000 signatures, leading to the expansion of protected forests.

Unfortunately, pro-environmental efforts surrounding the metagame are not well-defined and a framework for this pillar remains unclear. Few participants in my research discussed the metagame directly—those that did were unclear how the metagame might be applied in practice, despite their excitement about its potential. This lack of clarity highlights the need to develop more strategies to leverage the social aspects of gaming.

BUILDING A FRAMEWORK FOR CHANGE

It appears the overall aim of industry environmental advocates is to encourage pro-environmental behavior, or actions "which intentionally seek to reduce the negative impacts of a person's activities on the natural and built world."[29] This is commendable, but it's also very open-ended. What does it actually mean to "activate" and "impact" people? What specific "actions" should be sought, and why? What are the desired results? There seems to be little consensus on these sorts of questions in the GGM.

When considering the four pillars of action collectively, I argue that the GGM would benefit from a comprehensive and clearly defined *theory of change*. This involves determining which actions to pursue and establishing how these actions are expected to lead to some kind of change.[30] Some of this has already occurred, as evidenced by Playing for the

Planet's "theory of action" for game design[31] and the Sustainable Games Alliance's *Digital Game Environmental Standard*[32] aimed at reducing emissions in gaming. However, a broader, systematic framework that addresses all four pillars is needed.

Figure 11.2 depicts an extremely simplistic framework that might be used as a starting point for developing this theory of change. However, the theory of change should emerge from consultation with an array of stakeholders who are a part of the GGM. The movement is too complex for any single person or group to decide what is needed.

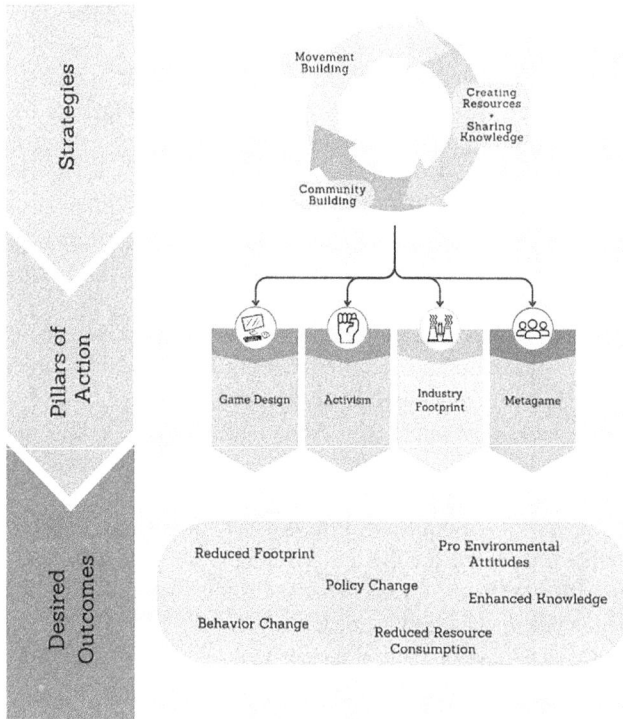

FIGURE 11.2   A framework for environmental advocacy in gaming.

(Source: Author).

Challenges to the SIG and GGM

Although developing a theory of change will be helpful, environmental advocates in the game industry still face two significant challenges.

Resource Deficiency

All participants in my research stressed that the lack of funding limits what the SIG can achieve. Andrew Brennwald confirmed this by stating, "It's difficult to cobble together our ambitions with the resources we have, which are very limited." Learning Designer and

Game Researcher Clayton Whittle noted how the SIG has reached a "critical mass," but the absence of funding is the key thing preventing the group from reaching its full potential.

The funding issue is linked to another challenge—a shortage of time. Because the participants have jobs and other commitments outside of the SIG, they cannot consistently contribute to the group's aims. This makes it difficult to retain members of leadership, which in turn affects engagement and organizational efforts. Some of this is likely due to the instability of the gaming industry and the emotionally draining nature of environmental advocacy. Nonetheless, it's challenging to keep unpaid volunteers committed to a cause.[33]

While any grassroots volunteer organization might face similar challenges, a lack of funding for environmental initiatives extends beyond the SIG and persists throughout the game industry. Specifically, there is a structural resistance to creating content outside of proven profitable paradigms,[34] which makes the game industry reluctant to invest in sustainable-themed games. Nick Schrag compared green games to "fighting an economic model," and Megan Carriker pointed out how difficult it is to "sell" the idea of them, especially from a publishing perspective. Some smaller studios are bucking this trend, but many large studios remain hesitant to produce pro-environmental content.

Green game design practices are not the only underfunded environmental initiatives in gaming. Clayton Whittle voiced his frustration regarding the lack of pro-environmental career pathways in the industry:

> There are so few companies that even have sustainability initiatives… It blows my mind that it is not top priority for game companies to have ESG officers and have sustainability be on the list of jobs that they need…[it's] what other industries already have.

The needle has begun to shift, and more studios are embracing sustainability roles. However, most industry workers are unhappy with game studios' pro-environmental efforts.[35] Additionally, indie studios are leading the charge while many large studios stand by. As Patrick Prax notes in the introduction, we need deeper, structural change, not greenwashing. This means more funding, more jobs, and a greater focus on sustainability within the industry.

Interdisciplinary Coordination

There is also a need for more collaboration among actors in the GGM. A persistent challenge is that game developers are attempting to address environmental issues but lack essential knowledge and skills. Conversely, advocates outside the game industry don't understand what it is like to interface with game-making practices. Clayton Whittle aptly described the problem:

> Everyone kind of thinks they already know how to do it. Right, the game devs think they already know how to do it. And the academics can make a game *(sarcastically)*. Hire a research assistant this year who knows how to program in C+, and they'll make it in Unity. And that'll be that.

Joost Vervoort summarizes this "rift" between those outside the industry and those within, stating, "Change makers don't understand games; game makers don't understand change."[36] In turn, Vervoort emphasizes the necessity of collaborating across these boundaries to foster change. Developers, academics, philanthropists, policymakers, NGOs, and artists should leverage one another's skills for good, rather than bicker about who is right or who can do it better.

There is also a lack of collaboration at the organizational level in the GGM. Jude Ower, co-founder of Playing for the Planet, noted this issue:

> there's lots of different groups doing different things… I find some more open than others. I don't know whether this is just because everyone's got different priorities… [but] there's no coordination between people who are leading in this movement.

Of course, fruitful alliances do exist, but there are not enough of them. It would also be helpful for partnering organizations to "formalize their relationship" (as Nick Schrag put it) so it's crystal clear how they can better support one another. If we can get hundreds or thousands of people lifting each other up and working together on a shared task, the future of the GGM will be bright.

TOWARD A BOLDER GGM

If you're reading this, you likely believe in gaming's potential to create positive change. You're probably excited that there are over 3 billion gamers. I bet you care deeply about games and want to see them play a role in addressing the pressing environmental issues of our time.

Unfortunately, at present, too much environmental advocacy in gaming is based on faith and hope. There is faith that games will impact players, and hope that this impact will make a difference. I recognize the value of these states of mind, and I am by no means disregarding them as trivial. However, to do our work effectively, we need more. We need a clear plan. This starts with a collaboratively developed, comprehensive theory of change. This will help organize pro-environmental efforts and begin to address the challenges facing the GGM.

This theory of change should extend beyond the linear model of: **Good Game Design + Player Interaction = Pro-environmental impact.** This paradigm is too limiting. Games are gathering places for people and ideas, even without direct environmental messaging. Solely focusing on design tactics narrows the possibilities for positive change.[37,38] It also places the impetus for action on players' shoulders. How can direct action be encouraged among developers? How can game communities be leveraged, both online and in person? How can we achieve a lower material footprint in the game production process? It's not enough to simply go toward sustainable game design; we must transcend it.[39]

One of the central tenets of the environmental movement is "organizing and motivating people to work towards a collective vision or cause."[40] As I've shown, the SIG is already a significant part of making this happen. The next step is bringing green game organizations and leaders together, and doing it consistently. When I worked for an environmental NGO in Colorado, we held a monthly call with representatives from organizations throughout the state. Not only was this a great way to share ideas and updates, but it was also a great opportunity for networking. Let's start something similar for the GGM. And if you aren't part of a GGM organization yet, join the SIG!

Pro-environmental initiatives in gaming often operate as side projects alongside other full-time work. This is a broken model. Work in the GGM should be built on paid positions. While the SIG is an incredible example of volunteer impact, it also highlights the limitations of such efforts.

When I asked Trevin York about funding barriers in the GGM, he said, "I think the money exists, and people say that they want to give this money. [The challenge is] untangling how the money more effectively reaches people." How do we address this in a financially precarious industry? Perhaps there is more room for reaching across disciplines to garner funding. Academia has done a remarkable job of exploring avenues in the GGM, and there is certainly nothing wrong with developing more partnerships between researchers and game developers. We also need more connections with philanthropists, policymakers, and NGO leaders—they have access to funding too.

The "STRATEGIES" project is an excellent example of a funded, interdisciplinary effort. Founded in 2024 by the European Union, STRATEGIES brings together a diverse array of European green game practitioners from academia, game development, and NGO sectors to tackle sustainable production, workplace practices, and game design.[41] The group has well-defined goals that address multiple pillars of the GGM and plans to develop resources and share information.

Where might we go from there? Consider what could happen if an interdisciplinary team of dedicated individuals versed in games, game culture, environmental issues, movement building, project management, marketing, and financial development worked together in full-time, paid roles. They could create games that explore the design practices discussed in this book. They could critically assess games and their place in society. They could explore games as rallying points for social movements and tools for mobilizing people who aren't gamers themselves. Importantly, the group could act as a conduit for connection within and beyond the GGM. Perhaps I'm daydreaming here, but I could see this group making a real impact.

I've made numerous suggestions in this chapter, many of which are lofty, time-consuming, or not applicable to many readers. I'll conclude with several simpler suggestions.

- Give talks, make videos, or start a streaming channel about green gaming.
- Become an organizer in the SIG.
- Start a group at work to foster dialogue about green gaming. Develop a strategy for sustainability and share it with your colleagues.
- Invite people to play a green game together. Even better, turn it into a regular event. Use it as a springboard for discussion and action.
- Find pro-environmental spaces outside of gaming to engage with. Join campaigns, attend events, etc., and share insights about the GGM with the people you meet.

This account of the GGM is not intended to be prescriptive. Some of my suggestions are already being implemented, while others may not be particularly useful. I certainly don't have all the answers, which is why I've posed so many questions. This rapidly growing movement demands diverse and evolving approaches, and there is no single, straightforward solution to enhance the effectiveness of our work. This is why we should challenge ourselves to think more broadly about the intersection of games and sustainability.

From the moment I began studying the GGM, I've been struck by the passion and dedication of those involved. The people in this movement are extremely adept at overcoming challenges while operating under demanding and time-sensitive conditions. Our ability to foster creativity is also unmatched. In many ways, this is a community perfectly suited to confronting environmental challenges. Now we must scale up our efforts and unite around our shared goals. There is much work to be done; why not start today?

NOTES

1 Rootes, Christopher. "Environmental Movements." In *The Blackwell Companion to Social Movements*, 608–640. Blackwell Publishing Ltd, 2004. https://onlinelibrary.wiley.com/doi/10.1002/9780470999103.ch26.
2 IGDA. "About Us," 2022. https://www.igdaclimatesig.org/about.
3 Chang, Alenda Y. "Change for Games: On Sustainable Design Patterns for the (Digital) Future." In *Ecogames*, edited by Laura op de Beke, Joost Raessens, Stefan Werning, and Gerald Farca, 73–88. Playful Perspectives on the Climate Crisis. Amsterdam University Press, 2024. https://doi.org/10.2307/jj.10819591.4.
4 Whittle, Clayton, Trevin York, Paula Escuadra, Grant Shonkwiler, Hugo Bille, Arnaud Fayolle, Benn McGregor, et al. *IGDA's Environmental Game Design Playbook*, 2022.
5 Chang, Alenda Y. *Playing Nature: Ecology in Video Games*. Minneapolis: University of Minnesota Press, 2019. https://muse.jhu.edu/pub/23/monograph/book/72358.
6 Pedercini, Paolo. "Making Games in a Fucked Up World – G4C 2014 – Molleindustria," April 29, 2014. https://www.molleindustria.org/blog/making-games-in-a-fucked-up-world-games-for-change-2014/.
7 Abraham, Benjamin, and Darshana Jayemanne. "Where Are All the Climate Change Games? Locating Digital Games' Response to Climate Change." *Transformations: Journal of Media, Culture & Technology* 2017, no. 30 (November 8, 2017): 74–94.
8 Vervoort, Joost M., Carien Moossdorff, and Kyle A. Thompson. "7. Games for Better Futures : The Art and Joy of Making and Unmaking Societies." In *Ecogames: Playful Perspectives on the Climate Crisis*, edited by Laura Beke, Joost Raessens, Stefan Werning, and Gerald Farca, 181–98. Amsterdam University Press, 2024.
9 Blake, Katie, Ugo Arbieu, Sofia Castelló y Tickell, Takahiro Kubo, Sandra Lai, Kota Mameno, Silvio Marchini, et al. "How Commercial Video Games Could Engage Players with Biodiversity Conservation: A Systematic Map Protocol." OSF, October 10, 2023. https://doi.org/10.31235/osf.io/gshdp.
10 Ustwo games. *Alba: A Wildlife Adventure*. Ustwo games. Playstation/Mobile. 2020.
11 ZA/UM. *Disco Elysium*. ZA/UM. Playstation/PC/Mac. 2019.
12 Vervoort, Joost. "Games, Activism and the Art of Destroying Shit." Medium, July 18, 2022. https://anticiplay.medium.com/games-activism-and-the-art-of-destroying-shit-8c2ab2d1dd1d.
13 Vervoort, Joost. "Game Designers Rebellion: Learning from the Resonant Realities of Climate Activism." *Medium* (blog), February 8, 2023. https://anticiplay.medium.com/game-designers-rebellion-learning-from-the-resonant-realities-of-climate-activism-e0ddae241285.
14 Abraham, Benjamin J. *Digital Games After Climate Change*. Palgrave Studies in Media and Environmental Communication. Cham: Springer International Publishing, 2022. https://doi.org/10.1007/978-3-030-91705-0.
15 Jayemane, Darshana. "Game Studies' Material Turn." *Westminster Papers in Communication and Culture* 9, no. 1 (June 13, 2017). https://doi.org/10.16997/wpcc.145.
16 Epic Games. *Fortnite*. Epic Games. PlayStation/Xbox/PC. 2017.
17 Ashton, Daniel. "Upgrading the Self: Technology and the Self in the Digital Games Perpetual Innovation Economy." *Convergence* 17, no. 3 (August 1, 2011): 307–321.

18 Dyer-Witheford, Nick, and Greig De Peuter. *Games of Empire: Global Capitalism and Video Games*. U of Minnesota Press, 2009.

19 Abraham, 2022.

20 Chang, Alenda, and John Parham. "Green Computer and Video Games: An Introduction." *Ecozon@: European Journal of Literature, Culture and Environment* 8, no. 2 (October 31, 2017): 1–17. https://doi.org/10.37536/ECOZONA.2017.8.2.1829.

21 Carter, Marcus, Martin Gibbs, and Mitchell Harrop. "Metagames, Paragames and Orthogames: A New Vocabulary." In *Proceedings of the International Conference on the Foundations of Digital Games*, 11–17.

22 Costiuc, Stanislav. "What Is A Meta-Game?," 2019. https://www.gamedeveloper.com/design/what-is-a-meta-game-.

23 Boluk, Stephanie, and Patrick Lemieux. "INTRODUCTION. Metagaming: Videogames and the Practice of Play." In *Metagaming*, 1–22. Playing, Competing, Spectating, Cheating, Trading, Making, and Breaking Videogames. University of Minnesota Press, 2017. https://doi.org/10.5749/j.ctt1n2ttjx.3.

24 Whittle et al., 2022.

25 Twitch Interactive. "Press Center." Press Center, 2024. https://www.twitch.tv/p/press-center/.

26 Speed, Abbie, Alycia Burnett, and Tom Robinson II. "Beyond the Game: Understanding Why People Enjoy Viewing Twitch." *Entertainment Computing* 45 (March 1, 2023): 100545. https://doi.org/10.1016/j.entcom.2022.100545.

27 Mojang Studios. *Minecraft*. Mojang Studios. PlayStation/Xbox/PC/Mobile/Switch/Mac. 2011.

28 Amadori, Gaia. "Gaming for Ecological Activism: A Multidimensional Model for Networks Articulated Through Video Games." *Games and Culture* 19, no. 5 (July 1, 2024): 551–570. https://doi.org/10.1177/15554120231170141.

29 Staddon, Sam C., Chandrika Cycil, Murray Goulden, Caroline Leygue, and Alexa Spence. "Intervening to Change Behaviour and Save Energy in the Workplace: A Systematic Review of Available Evidence." *Energy Research & Social Science* 17 (July 1, 2016): 30–51. https://doi.org/10.1016/j.erss.2016.03.027.

30 United Nations Development Group. "THEORY OF CHANGE UNDAF COMPANION GUIDANCE," 2017. https://unsdg.un.org/resources/theory-change-undaf-companion-guidance.

31 *P4P Webinar #18: Defining a Theory of Action: How Do to Set and Measure Outcomes, 2021.* https://www.youtube.com/watch?v=9Zse-cGr3RI.

32 Sustainable Games Alliance. "Sustainable Games Alliance Digital Game Environmental Standard Draft v0.2," 2024. https://sustainablegamesalliance.org/standard/read-the-standard/.

33 Sextus, Charlotte P., Karen F. Hytten, and Paul Perry. "A Systematic Review of Environmental Volunteer Motivations." *Society & Natural Resources*, November 1, 2024. https://www.tandfonline.com/doi/abs/10.1080/08941920.2024.2381202.

34 Paul, Christopher A. *The Toxic Meritocracy of Video Games*. University of Minnesota Press, 2018. https://doi.org/10.5749/j.ctt2204rbz.

35 IGDA. "Developer Satisfaction Survey 2023 Summary Report," 2024. https://igda.org/dss/.

36 Vervoort, Joost. "Change Makers Don't Understand Games; Game Makers Don't Understand Change." *Medium* (blog), June 5, 2024. https://anticiplay.medium.com/change-makers-dont-understand-games-game-makers-don-t-understand-change-65834a420747.

37 Abraham, 2022.

38 Pedercini, 2014.

39 Apologies to the editors.

40 The Solutions Project. "Build a Movement." The Solutions Project, 2025. https://thesolutionsproject.org/info/build-a-movement/.

41 STRATEGIES. "About." Accessed November 21, 2024. https://www.strategieshorizon.eu/about.

Exploring Sustainable Gamemaking through an Artgame Residency

12

Cindy Poremba, Kara Stone,
Ian Garrett, and Ben Abraham

ABSTRACT

The environmental impact of video games has been receiving increasing attention, but there is little focus on the role of practice in creating low-carbon games. In what ways can game creators consciously shift the way they make games toward more sustainable and climate-conscious models? With this question in mind, the first Renewable Artgame Residency ran in Summer 2023. It offered an extended format that could be described as a game jam/artist residency hybrid, where participants lived and created games exclusively using renewable solar energy. This chapter will give an overview of the residency's values-led design, and insights from the consultation and evaluation process, including a reflection on different carbon-reduction and renewable energy approaches at the scale of indie and/or artgame creation.

KEY TAKEAWAYS

- **Game dev has a carbon cost.** Despite their digital nature, games rely on energy-hungry infrastructure that creators often overlook.

DOI: 10.1201/9781003512400-14

- **Practice matters.** The Renewable Artgame Residency showed that rethinking *how* games are made can be just as important as *what* they say about climate.
- **Off-grid isn't the only goal.** Conscious energy use, slower workflows, and reflective design offer more sustainable alternatives to crunch culture.
- **Small games, big lessons.** Indie and artgames may have lower emissions, but they can model powerful climate-conscious practices for the wider industry.
- **Residencies can spark change.** Values-led, renewable-energy-focused residencies empower creators to align their process with their environmental values—and reimagine the future of sustainable gamemaking.

There is a growing body of journalism and scholarship exposing the environmental impact of video games, from the resource consumption of the developers creating them to the energy costs of the computers playing and/or streaming them. Yet, few discussions focus on the role of design practice itself in creating low-carbon games. How can an artist, or other small-scale creator, meaningfully shift their gamemaking practice toward more climate-conscious, sustainable models?

Funded in part by the Canada Council for the Arts' strategic *Digital Greenhouse* program, the *Renewable Artgame Residency* (RAR) was a prototype initiative, run in Summer 2023, by (co-authors) gamemakers Cindy Poremba and Kara Stone. It focused on low-carbon digital design methods and new modes for sustainability-focused creation practices, specifically targeting smaller scale (art, experimental, and indie game) creation practices. This experimental initiative looked to reorient game creators toward the impacts of digital game creation as part of an expanded ecology that includes hardware and material resources, the energy cost of computing and its intersection with common development and creative practices, the impact of creative outputs (including distribution and gameplay), and, in the context of these challenges, the personal impact of the creative process. We wanted to look beyond "greenwashing" approaches to suggest effective solutions and models, developed with and through practice, that could both inspire and support game creators who hope to engage in more sustainable creative practices.

The RAR had three primary goals: (1) to explore different modes of slower paced, reflective game design; (2) to envision values-led sustainable practice in the context of artist-created games, and (3) to provide space for more environmentally responsible digital use and creation. The residency design was informed by consultations from (co-author) Benjamin Abraham, an expert in the environmental impact of gamemaking, and (co-author) Ian Garrett, who provided guidance on the use of renewable energy infrastructures in media artworks. This consultation process helped structure the parameters of the residency and guide evaluation criteria suited to the small-scale practices of artgame making. The residency was further reviewed by an external examiner (Apihtawikosisân game designer Meagan Byrne), who offered a participant perspective outside of the residency design team. As our goal was to examine game creation practices, our methodology centered on reflective practice occurring both on an individual level and collaboratively through knowledge, insight, and creative practice sharing over the course of the residency itself. It was followed by a postmortem into the experience, including both reflection on practice and on the efficacy of the residency model. Outcomes from the residency were then shared through multiple channels, including a solar-powered website and a RAR zine.[1] This chapter shares some of our insights into the role residencies might play in a greater shift toward sustainable gamemaking practices.

THE ENVIRONMENTAL IMPACT
OF VIDEO GAMES

The negative environmental impact of video games mirrors that of general-purpose computing and digital media, with the added sting of video game creation and gameplay being leisure activities (in contrast to essential digital services like, say, banking, or computer-aided design and engineering). Documented and known impacts of video game creation and consumption include everything from resource depletion in the process of raw material extraction for gaming devices,[2] to the energy consumed by hundreds of thousands of game developers toiling away in offices around the world,[3] to the fossil-fuel-based electricity consumed in internet backbone infrastructure and edge networks.[4] Mountains of plastics discs and other physical media are still sent all around the world to distribute games,[5] and lastly—and perhaps most substantially—the end users themselves generate negative impacts through yet more fossil-fuel-based electricity consumed by the billions of player around the world,[6] and contribute to municipal waste disposal burdens when digital devices pass into obsolescence,[7] rarely if ever recycled[8] (for an expanded discussion of these issues, see Chapter 4).

Despite this, digital game creation continues to maintain a veneer of environmental neutrality. While environmental awareness may be growing, artists working within this "virtual" space can still do so without consciously engaging with the environmental and/or climate impacts of gamemaking, which conveniently occur indirectly and in far-flung sacrificial zones elsewhere from the location of both production and consumption. Creators who do wish to be more mindful of environmental and climate concerns have limited models or resources from which to draw from.

The game industry, to date, has primarily focused its environmental initiatives on communicating climate impact and sustainability to players. A major focus of the first climate-oriented games organization (the *Playing for the Planet Alliance*[9]) points toward the potential to engage the billion gamers around the world with climate activations or awareness, scale being both a feature of the game industry itself and, consequently, a necessity for effective climate action. The IGDA Climate Special Interest Group has produced an *Environmental Game Design Playbook*[10] with design strategies for integrating more effective sustainability messaging within games. And game companies have released environmentally themed content for both new and franchise titles (for example, Electronic Arts' 2020 "Eco Lifestyle" expansion for its popular *Sims 4* title). These initiatives are one important strategy to bring increased awareness to environmental issues, educate players on the complex systems at play in the global climate crisis, and reorient players toward eco-positive perspectives and behaviors.

Researchers and gamemakers are just beginning to address the carbon impact of *game development* itself,[11] and much of this work focuses on larger scale (commercial studio) development practices and outputs. Looking at Electronic Arts again, we can see the company report on significant investments in renewable energy and efficiency[12] toward their production pipeline. Guidance for small-scale gamemakers has lagged behind these large-scale players, although there are notable exceptions: for example, Benjamin Abraham's *AfterClimate* consultancy has conducted climate impact audits for both Die Gut Fabrik's *Saltsea Chronicles* (2023) and Kara Stone's own solar game server. Recognizing and renewing the impact of both large and smaller scale development practices remains important and necessary, as a strong multi-threaded approach from all cultural creators is perhaps the most effective avenue for impact. While not every approach adopted by artgame creators will scale, such approaches can still drive cultural leadership and offer models and insights that can inform larger studio and industry practices.

BEYOND GAMES: RENEWABLE ENERGY AND MEDIA ART

Artgame creation bridges two adjacent but distinct cultures: game development and media art. As part of our consultation process, we looked beyond game development to see what media art may offer in terms of opportunities and challenges when it came to reorienting practice toward sustainability.

In the wider field of media art, we found the same ecological concerns (and many of the same approaches to mitigating negative environmental impacts) are at play. Modes of content and asset creation can be heavily influenced by gamemaking, and media artists commonly work in the same technocultural environments as gamemakers. However, there are key differences as well. One is the degree to which an audience "plays" the work. In video games, we have begun to recognize the environmental costs that may emerge simply from the scale of large numbers of players playing computationally intensive games on high-end systems, where we seldom see that intensity of participant activity with media artworks. Another is the context in which the work is shown: unlike most commercial games, which are experienced on personal devices, media artworks are commonly experienced as events, often through shows, exhibitions, and/or festivals. This potentially means they have a more concentrated carbon footprint, but for a shorter duration and using fewer individual systems. Further, the experimental charge of media art can open opportunities to challenge existing energy conventions, and explore different ways of creating or showing energy-driven works, without the pressures of mass-consumer expectations. These differing contexts mean that there are some unique considerations for works that might bridge media art and video game spaces.

Media artists have notably embraced the potential, and unique affordances, of renewable energy sources. Alex Nathanson (a New York-based artist, scholar, and author of *A History of Solar Power Art and Design*[13]) has created a variety of projects integrating solar power, including *Solar Power Drawing Machine Studies*[14] which uses small solar panels to power motorized drawing implements. In the work, Nathanson is able to use the variability of solar power as a creative tool, having patterns emerge as the devices shift their relationship to their light source, sometimes stalling or following meandering paths. Another example, the 2015 dance performance by Zata Omm Dance Project, *vox:lumen*,[15] was initially designed to use kinetic power generation, using the movement of the performers (through dance) and audience (through bike generators) to generate electricity. However, this approach to power generation created issues with both the consistency of the power flowing in, and placed constraints on the performance itself, which eventually led the team to pivot to solar.[16] These examples of the use of renewable power highlight the challenges of working with the affordances of variable technologies in mind—even before dealing with the introduction of gameplay. But they also direct our attention toward some of the affordances of more energy-conscious approaches. What if we stopped looking at sustainable practices as being solely about constraints and embraced the quirks and alternate flows of renewable energy?

RELEVANCE FOR SMALL-SCALE GAMEMAKERS

Within this larger picture of environmental considerations, what actually matters for indie and artgame creators looking to cultivate more sustainable practices? Working toward different outcomes (for example, different audiences and/or different play contexts), and at a much smaller scale, it can become challenging to tease out what meaningful climate action might look like. While it is essential to address known issues like the environmental costs of game and media streaming, large-scale and uncontrolled power use from a player base, development impacts from AI/machine learning, and large-scale data processing, these issues might not directly intersect with artgame practices. These practices commonly leave much smaller development footprints, relying on smaller teams with less resources, and are aimed at much smaller player bases (including gallery audiences, festival-goers, and individual players accessing the work via online downloads). Art games additionally tend to be shorter (in part because of the limits of their development resources, and in part because their reception context often presumes a reduced timeframe for engagement). As such, even art games that lie on the far end of the high-fidelity, high-resource production end of the spectrum, simply move around less data and consume fewer resources.

This may be one reason why indie and artgame creators have engaged issues of sustainability primarily through game *themes*, both ambiently and via persuasive games.[17] As Alenda Y. Chang has suggested, we can and should pursue eco-critical games as one move toward more climate-responsible action.[18] Eco-critical independent games like *Half-Earth Socialism*[19] and *Green New Deal Simulator*,[20] for example, model complex environmental dynamics, offering players insight into potential paths forward. Others, like 2022's *Norco*[21] and Tali Faulkner's *Umurangi Generation*,[22] house broader narratives within visions of dystopian climate futures. Still, not only are eco-critical games underrepresented within the broader spectrum of games with environmental thematics, many such games still perpetuate problematic ecological relations and/or are contained within the sub-genre of educational games.[23]

There are fewer examples that move beyond ecological, sustainability, and/or environmental game themes, into eco-critical critiques of the materiality of video game creation and gameplay. Angelo Vermeulen's 2007 *Biomodd*[24] highlighted the excess waste heat produced by high-end computer systems, by creating ecosystems living within these cases. More recently, Alex Custodio and Michael Iantorno, as part of the Solar Media Collective,[25] have created workshops around solar game boy modding and homebrew game design. These workshops not only foreground more sustainable power integration into game hardware, but also critique the designed obsolescence and relentless upgrade cycles of said hardware, by reclaiming older hardware as viable development platforms. Co-author Stone has also created a solar-powered web server that hosts video games designed to be low-carbon, tackling how video games can be shared with players without relying on the consumption of fossil fuels.

Despite this important work, game artists taking up climate-conscious approaches remain oddly few and far between (particularly given the existential threat of the climate crisis). And while gamemakers are bringing more attention to environmental themes, and gaming hardware and infrastructures, there remains a gap in considering gamemaking practices themselves. Could long-form game jams and artist residencies serve as an intervention point, one where artists could create low-carbon games, learn and share relevant environmental practices, and renew not just what they were making but how they were making it?

DESIGNING THE RAR

At its heart, the RAR questions the norms and values embedded in our game creation practices. The typical modes of creative practice in the field of games are rapid-paced iteration, crunch, and game jams,[26] where creators are often pressured to create work quickly, without much time for reflection, reconsideration, and revaluation. This dynamic extends from the commercial game industry into artgame making, despite the potential for slower, more mindful creative practice.[27]

It's important to stress that there is nothing innately wrong with using existing models such as game jams[28] to address environmental issues. But given both our desire for more reflective practice, and our target audience of small-scale game creators (makers who commonly span both art and game cultures), we felt that something more akin to an artist residency, situating individually sustainable work patterns alongside environmental sustainability, was both an underexplored and promising model.

Artist residencies are common in the fields of fine art, media art, and writing, and have long been used to disrupt and/or renew creative practice, including in eco-critical contexts.[29] Their purpose is to afford dedicated time and energy to practice, in part by removing the participant from other day-to-day commitments, routines, and distractions. As such, they typically carve out both time (anywhere from a week to a year) and space (commonly offering the artist living and/or studio space), in a new environment that might bring a renewed creative energy or focus. Residencies may also offer community and networking opportunities with other artists, allowing for an exchange of new ideas and ways of working.

Residencies normally require travel, and this can be a particular challenge for artists based in North America, where transportation infrastructures offer fewer low-carbon options. One of the first design concerns of the RAR was to imagine ways we could afford the necessary *apartness* that would support focus, reflection, and a shift in practice, while reducing the potential impact of travel. But we wanted to do this without also impacting the *togetherness* dynamic of residencies, where the presence and support of other creators might offer an opportunity to cultivate a community of practice. To address this, we trialed a "remote-together" model. Each participant secured an off-grid accommodation local to them, and we provided funding for these local accommodations, plus a solar generator, and a per diem to cover food for the two-week duration. This distributed model offset the biggest carbon cost associated with the artist residency—transit—but meant that each participant found their own accommodation and was on their own during the residency period. To foster the network and shared experience of the residency, even while not physically together, we set a virtual meeting schedule. Participants agreed upon a time to check-in using voice, text, or telephone chat on Discord each day to share progress synchronously, and stayed loosely connected at other times asynchronously. This included sharing progress on game creation, tactics for working with solar generators and off-grid setups, reflective thoughts on the experience, as well as general daily life updates. This allowed for a discovery-based approach to working more sustainably, where participants were not only able to explore self-directed approaches but also to share successes and challenges.

Another early question that emerged in the residency prototype was whether, and/or to what extent, we would mandate that the games deal with renewable energy, or environmental *themes*. Abraham has noted that if we are to strategically think about maximizing the impact of sustainable practices in game creation, we should be looking to make sustainable practices the norm for *all* games, not just ones with environmental themes.[30] If eco-games are only ever eco-themed, to some extent, we miss out on an opportunity to normalize sustainable practices across the board. As a counterpoint, themes are a common, and often expected, part of both game jams and residencies. They typically create cohesion between a group of participants that can help creators come together and potentially find resonance between projects. As part of our RAR prototype, we opted not to demand a strict adherence to a theme, but instead suggested that we think about the environment as a whole. That loose suggestion resulted in several different game approaches with varying degrees of connectedness to eco-critical themes.

Part of the practical planning for the RAR required thoughtful consideration of equipment and infrastructure to enable sustainable, off-grid creative practices. Portable solar-power stations were a key element of the residency's design. These solar generators powered the laptops used for gamemaking, the phones used to communicate and access the internet (where possible), and general power needs over the course of the residency. Pragmatically speaking, they needed to be portable enough to transport to the residency location, while also having the capacity and flexibility for participants with intensive energy demands without being overly restrictive. The recharge capabilities and speed of input from the necessary accompanying solar panels were an additional consideration in planning the residency's renewable energy approach.[31]

Logistics for the residency prototype extended beyond selecting equipment. Weather patterns, solar exposure, and physical infrastructure need to be assessed to ensure successful setups. Participants needed to be guided on managing energy usage, scheduling tasks around peak solar generation, and preparing for limited battery reserves during overcast weather. These considerations helped balance renewable energy constraints with the demands of creative work, fostering an awareness of energy use often absent in traditional residencies. Most importantly, by taking power off the "free-running tap" of the public energy grid, we encouraged participants to attune to energy use and to develop practices that allowed for recharging, integrated breaks, and discouraged thoughtless consumption. The use of solar energy may have had a minor material impact (depending on where the participants' regular practice was located), but a substantive impact on how game creation attended to energy flow and consumption.

RESIDENCY OUTCOMES

The prototype for the RAR focused on individual creators looking to orient themselves toward a more ethical sustainable practice. Even with this focus, we anticipated gamemakers would come to the residency for different reasons: to have a direct environmental impact no matter how small, to relieve or explore personal climate anxiety, or perhaps to feel authentic to one's lived values. As such, we needed to consider how we would measure and evaluate the impact of our approach. Where we initially thought the residency would allow us to offer a clear set of guidelines to offer small-scale gamemakers surrounding sustainable small-scale creation practices, the reality is that the potential outcomes from this model are wider ranging and less generalizable than we had expected in our initial planning. To adjust our evaluation strategy and output, we decided to embrace "softer data" emerging from the prototype and highlight insights that emerged through the residency development.

One priority for reflection and evaluation was our residency model. With a distributed residency setup, we were successful in reducing carbon emissions from air travel that would be used in a typical artist residency, while maintaining the desired qualities of both *apartness* and *togetherness*. However, we also concluded that "off-grid" was not, by default, meaningful climate action. For example, one participant had challenges finding off-grid accommodation that was supported by renewable power, with most off-grid opportunities in their area powered by non-renewable sources like propane gas. Another participant opted to engage in voluntary off-grid practices in a location that was technically on the power grid, to ensure they had a manageable travel plan and had fallback options to manage their disability. These issues highlighted just how tangled our relationships with renewable energy can be, and the need to balance the value of intentional resource constraints with pragmatic considerations surrounding disconnecting from a reliable power grid. The remote nature of the selected RAR locations also meant that all participants at some point needed to manage access to the internet (for daily check-ins), as well as adapt to extreme weather and climate challenges.[32] What emerged as most important in relation to the off-grid criteria was an awareness of energy-as-resource, and the associated adjustment of working practices. However, it's important to consider that different models and considerations exist, in terms of access and ability, that make participant choice a key part of any residency design.

In terms of the individual sustainable work patterns that the RAR aimed to provide time and space for, most participants found that it took at least a few days to switch into a pace and mode of work that felt manageable. Participants self-observed an internalized pressure to produce quickly (in line with a game jam model), that did not fit within the structure or goals of the RAR. Still, over the course of the residency, participants necessarily recalibrated around the constraints and affordances of working off-grid: for example, physically followed the sun around to ensure the generators were charged, careful not to deplete them after sunset or with forecasted rain. While the two-week timeframe was initially chosen to better accommodate working gamemakers, in the end, two weeks felt too short to fully adjust to new accommodations, new technology, and new work rhythms. Noticing and identifying these kinds of points of friction was a valuable step toward reimagining individual practices and patterns.

Each participant came into the residency with personal goals related to sustainability and their gamemaking practice. For one participant, the residency was an opportunity to bring carbon-consciousness into the creation portion of their game design work, alongside their existing eco-critical game design, and efforts to find more sustainable distribution platforms. For them, the residency became a way to align their development practices with their broader environmental ethics, to create a more wholistic approach to sustainable game development. For another, the residency was part of a process of burnout recovery—an opportunity to find new ways to pull positive energy from the creative practice of game design. Their focus during the residency was on re-embracing analog design practices. Another participant aimed to reconnect with more lightweight technologies, including recovering use of an old laptop, and re-engage over-compression aesthetics and web coding outside of major game development engines. Their approach embraced the idea of "graceful decline,"[33] reverting their normally computationally intensive gamemaking approach to a more minimal, and more mindful, model.

With no constraints on approach or outcomes, and no mandate to even create work that was explicitly eco-critical in theme, the participant outcomes varied in interesting ways. For example, while not explicitly climate-themed, one game concept drew from moss photos gathered while wandering the forest, and used data from a live weather API to have in-game technologies also run via solar power. Inspired by how the residency was able to attune them to power as a managed resource, they were able to translate this insight into game mechanics with a direct link to real-world environmental conditions. Another participant went back to pen-and-paper to prototype mechanics around two new creative projects. Rather than jumping directly into digital prototyping, resisting this impulse made room for them to spend more time simply exploring their design process, and finding pleasure in sketching, sharing concepts, and playing with game ideas. Even without explicit directives to create a particular type of work, the re-imagined working structure had ripple effects extending into what was created, and how it was created—just not necessarily in the ways we may have initially expected. But this again returns us to one of the key insights of the residency: that it is essential for artgame creators to maintain the freedom to determine what a successful outcome looks like, rather than have the residency attempt to funnel creative practice toward a pre-determined, generalized vision of sustainable gamemaking.

In fact, what we have found is that successful outcomes around shifting creative practices to be more personal and life-sustainable can be highly individual and contextual, which makes generic recommendations and metrics less helpful. This runs counter to our broader understanding of eco-recommendations and tracked metrics as being extremely beneficial for improving environmental outcomes. This tension speaks to broader issues well beyond the scope of an artgame residency prototype. Nevertheless, our early work suggests that providing a structure for creators to develop their own sustainable practices, and from there building out a larger community of practice, can be a promising alternative model for cultivating sustainable game development.

NOTES FOR A RENEWABLE RESIDENCY

Gathered from this initial prototype of the *Renewable Artgames Residency*, we share the following recommendations for creating game-centered artist residencies focused on sustainability and/or renewable energy:

1) Consider duration, ideally extending to several weeks. While longer form residencies/game jams require an increased commitment, the opportunity presented by a RAR is in finding new ways of engaging in game creation practices, and this lived experience requires an investment of time.
2) Networked local accommodations (from camping to off-grid to a renewable-energy-powered house) can be a workable model for reducing transit-related carbon costs.
3) Structure a schedule for virtual communication between the participants and the organizers that will: (a) work in remote locations and (b) allow for varying levels of disconnection (both intentional and sporadic).
4) Focus on reflection rather than output. While a reflection into one's own practice can be deeply individual, it can be useful to have a mechanism to have this reflection on practice made tangible or shareable in some way: whether that be sharing personal reflections within a residency cohort or back in some other way to a community of practice (for example, a public or studio talk, or even a piece of reflective writing on the experience).

FINAL THOUGHTS

Video games exist within a larger landscape of media art where artists and designers have been creatively addressing the climate crisis and integrating renewable energy into the material practices of creation. Though the majority of the carbon emissions come from AAA gaming, small-scale game design practices can also help prompt practical and beneficial impact.

The RAR offers an initial model for reimaging game jams and residencies within a sustainability framework. By emphasizing slower paces, values-led approaches, creative residencies like RAR can give space for designers and developers to combine digital games with environmental concerns and climate action strategies without the pressures of rapid iteration and crunch. The residency's structure, aimed at individual and collaborative reflection, prompted participants to rethink their engagement with work practices and energy usage. The focus on renewing development practices can help create alternatives to less sustainable practices in modern gamemaking and find ways of envisioning video games within the context of a climate-conscious future.

Building on the foundation laid by the RAR, the principles of energy-efficient design and sustainability can extend to broader contexts within the gaming industry and educational frameworks. For larger studios, experimenting with sustainability benchmarks in production pipelines offers a tangible way to reduce environmental impact while fostering innovation. Integrating residencies focused on renewable energy and resource constraints could cultivate a culture of climate-conscious production, enabling developers to explore creative solutions while addressing real-world challenges. These practices not only align with the industry's growing accountability toward environmental sustainability but also demonstrate leadership in advancing eco-conscious methodologies.

In academic settings, sustainability should become a cornerstone of game design pedagogy, equipping future creators with the knowledge and tools to address environmental issues through their work. Assignments or capstone projects requiring the use of renewable energy sources or adherence to strict resource constraints, as we attempted to model through the RAR, can simulate the complexities of sustainable creation. Moreover, instructors can guide students to explore the integration of eco-friendly systems and renewable energy into gameplay mechanics and narratives, fostering a holistic understanding of how sustainability intersects with creativity. Such approaches can inspire emerging designers to envision games not just as entertainment but as powerful vehicles for ecological awareness and action.

As the gaming industry and its adjacent fields continue to grapple with their environmental impacts, initiatives like RAR serve as a call to action for both game companies and creators. By embedding sustainability into the fabric of game development, the industry can reimagine its relationship with energy use, resource consumption, and climate narratives. This shift not only aligns with the urgent need for climate action but also positions games as a medium capable of leading cultural and technological change in a climate-conscious world.

NOTES

1 Renewable Artgame Residency Zine https://www.solarserver.games/files/RARZine_digital.pdf.

2 Abraham, Benjamin. *Digital Games after Climate Change*. Palgrave Studies in Media and Environmental Communication Ser. Cham, Switzerland: Springer, 2022.

3 Abraham, 2022.

4 Marsden, Matthew, Mike Hazas, and Matthew Broadbent. "From One Edge to the Other: Exploring Gaming's Rising Presence on the Network." *Proceedings of the 7th International Conference on ICT for Sustainability*, 2020, 247–254.

5 Maxwell, Richard, and Toby Miller. "'Warm and Stuff y': The Ecological Impact of Electronic Games." In *The Video Game Industry*, edited by Peter Zackariasson and Timothy Wilson. Routledge, 2012.

6 Aslan, Joshua. "Climate Change Implications of Gaming Products and Services." *Doctoral Thesis, University of Surrey*, 2020. https://openresearch.surrey.ac.uk/esploro/outputs/doctoral/99512335802346.

7 Gabrys, Jennifer. *Digital Rubbish*. University of Michigan Press. 2011.

8 Becker, *Insolvent*; Hazas and Nathan, "Introduction"; Comber and Eriksson, "Computing as Ecocide."

9 Playing For The Planet. https://www.playing4theplanet.org/

10 Whittle, C., York, T., Escuadra, P.A., Shonkwiler, G., Bille, H., Fayolle, A., McGregor, B., Hayes, S., Knight, F., Wills, A., Chang, A., & Fernández Galeote, D. "Environmental Game Design Playbook (Presented by IGDA Climate Special Interest Group)." *International Game Developers Association*, 2022. https://igda.org/resources-archive/environmental-game-design-playbook-presented-by-igda-climate-special-interest-group-alpha-release/.

11 Klammer, "Videogames for Future"; Abraham, *Digital Games after Climate Change*; Werning, "Ecomodding. Understanding and Communicating the Climate Crisis by Co-Creating Commercial Video Games"; Whittle, C., York, T., Escuadra, P.A., Shonkwiler, G., Bille, H., Fayolle, A., McGregor, B., Hayes, S., Knight, F., Wills, A., Chang, A., & Fernández Galeote, D., "Environmental Game Design Playbook (Presented by IGDA Climate Special Interest Group)."

12 According to EA's 2024 impact report, 93% of their operations use renewable energy, 30% of it directly through on-site solar and geothermal power. They've also developed a system of Known Version Patching, which has reduced update patch size by 80% in part to reduce the energy required to push updates.

13 Nathanson, Alex. *A History of Solar Power Art and Design*. 1st edition. New York London: Routledge, 2023.

14 Nathanson, Alex. *Solar Power Drawing Machine Studies*. 2024. Solar power drawing machine. https://alexnathanson.com/projects/pvdrawing.html.

15 Yong, William. *Vox:Lumen*. March 2015. Dance. https://www.youtube.com/watch?v=5P9uHWLmHpE.

16 In contrast to Nathanson, solar is here used to bring a more consistent and stable flow of power to the work.

17 Abraham and Jayemanne, "Where Are All the Climate Change Games?"; De Beke, Raessens, and Werning, *Ecogames*; Chang, *Playing Nature*.

18 Chang, Alenda Y. *Playing Nature: Ecology in Video Games*. Electronic Mediations 58. Minneapolis: University of Minnesota Press, 2019.

19 Tseng, Francis, and Son La Pham. *Half-Earth Socialism*. 2022. Computer game. https://store.steampowered.com/app/2071530/HalfEarth_Socialism/.

20 Molleindustria. *Green New Deal Simulator*. 2023. Computer game. https://www.molleindustria.org/GND/.

21 Geography of Robots. *Norco*. 2022. Computer Game.

22 Origame Digital. *Umurangi Generation*. 2020. Computer Game.

23 Abraham, B. J., and D. Jayemanne. "Where Are All the Climate Change Games? Locating Digital Games' Response to Climate Change," November 8, 2017. https://opus.lib.uts.edu.au/handle/10453/121664.

24 Vermeulen. *Biomodd*. 2007. Installation.

25 Solar Media Collective. https://www.solar-media.net/.

26 Kultima, "Negative Game Jam Experiences"; Borg et al., "Video Game Development in a Rush."

27 Stone, Kara. "Reparative Game Creation: Designing For and With Psychosocial Disability." *Design Issues* 39, no. 1 (January 1, 2023): 14–26. https://doi.org/10.1162/desi_a_00703.

28 Since 2020, *Playing the Planet*, for example, runs an annual Green Game Jam in partnership with the United Nations Environment Programme; and local organizations, such as the *Nordic Alliance for Sustainability in Gaming*, have run multiple successful eco-game jams (see Wirman et al.).

29 LeBel, Sabine. "Facilitating Queer Art in the Climate Crisis." In *Facilitating Community Research for Social Change*, by Casey Burkholder, Funké Aladejebi, and Joshua Schwab-Cartas, 195–207, 1st ed. London: Routledge, 2022. https://doi.org/10.4324/9781003199236-17.

30 Abraham, *2022*.

31 Based on the experience with renewables in arts settings, Ian Garrett pointed to a variety of options. This included the Bluetti EB70S, EcoFlow Delta 2, and Jackery Explorer 3000 Pro. Each of these options were considered when thinking about the participants' needs. In this case, the smaller systems like the Bluetti EB70S offered portability and affordability for basic tasks, while larger models like the EcoFlow Delta 2 and Jackery Explorer 3000 Pro provided greater capacity and flexibility for more intensive energy demands. Note that these equipment considerations reflect both the year (2023) of the residency prototype and the constraints of the project funding.

32 Notably, the RAR took place in 2023 during what is sure to be one of many intense fire seasons in Canada, with numerous large-scale fires that cast thick smoke and heavy particulate mass blanketing major North American cities like Toronto and New York.

33 Marks, Laura. "Collapse Informatics and the Environmental Impact of Information and Communication Technologies." In *The Routledge Handbook of Ecomedia Studies*, edited by Adrian Ivakhiv, Alenda Y. Chang, Antonio López, Kiu-wai Chu, Miriam Tola, and Stephen Rust, 119–128, 2023. https://doi.org/10.4324/9781003176497-14.

Changing the Question

13

What Is an Ecological Game? On the Three Definitions of Sustainable Game Design

David Harold ten Cate

ABSTRACT

There are three ways of talking about the sustainability of videogames. The first mode considers games as media texts or cultural artifacts that can communicate about issues related to the climate crisis or the environment. The second approach sees games as environmental artifacts that rely on circuits of energy and material resources to exist. The third perspective regards games as social artifacts, representative of the constraints of game development. This chapter allows game developers to reflect on the balancing act involved in making their game sustainable. It offers some practical implications for the conceptualization of ecogame design.

KEY TAKEAWAYS

- **Sustainability is more than themes.** Games must be designed not just *about* the climate crisis, but *within* it—across narrative, production, and labor practices.

DOI: 10.1201/9781003512400-15

- **Every game leaves a footprint.** From hardware waste to energy use, developers must account for the material and environmental cost of game creation and play.
- **People make games—treat them sustainably.** Industry labor practices, from crunch to layoffs, must be part of any definition of ecological game design.
- **Resist the myth of apolitical games.** Climate-conscious design means addressing power, place, and responsibility—not escaping them through neutrality.
- **There is no quick fix.** Real sustainability demands rejecting harmful trends like GenAI and console supersessionism in favor of long-term, community-rooted change.

INTRODUCTION

The climate crisis has arrived, and its effects are already palpable.[1] Increasingly, alarming signals of impending catastrophe are being communicated.[2] The videogames industry finds itself in the middle of this crisis. Naturally, game developers are cognizant of the state of the world and may look to attune their practice to the mitigation of the climate crisis. For them to do this effectively and with an eye to systemic and material change as discussed in the introduction of this book, it is necessary to take a conceptual step back before taking the necessary steps forward. This chapter situates the question of sustainable game design in three definitions (Figure 13.1). This chapter also thinks about how these categories relate to each other. This conversation is important because it emphasizes that sustainable game design is necessarily a balancing act between different ways of being sustainable.

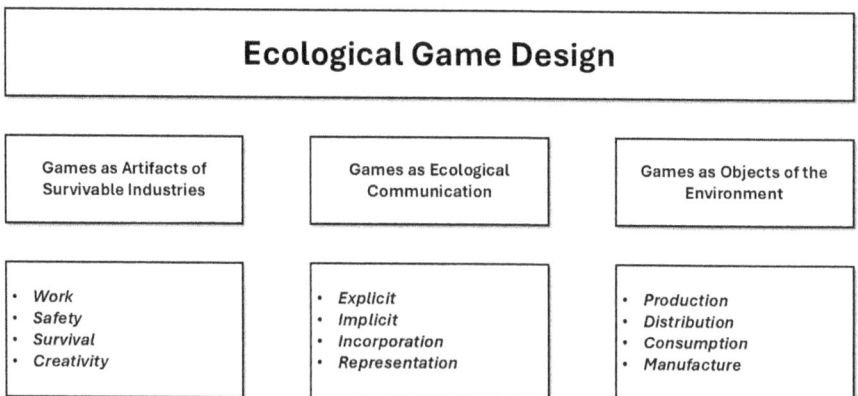

Ecological Game Design

Games as Artifacts of Survivable Industries	Games as Ecological Communication	Games as Objects of the Environment
• *Work* • *Safety* • *Survival* • *Creativity*	• *Explicit* • *Implicit* • *Incorporation* • *Representation*	• *Production* • *Distribution* • *Consumption* • *Manufacture*

FIGURE 13.1   The three ecologies of game design. Original by the author.

As represented in Figure 13.1, this chapter presents three interconnected definitions of *ecological games* (*ecogames*). At times, ecogame design may be dedicated explicitly to one form of sustainability and thereby sacrifice the other—equally important—areas of sustainability. An approach that connects these different forms is therefore more productive. The primary purpose of this chapter is to allow game developers to map the ecological relations of their game. The secondary purpose of this chapter is to provide some practical insights into the choices that may make games more sustainable across these categories.

Ecogame Definition 1: Games as Ecological Communication

The first definition of sustainability in game design concerns games as media texts. According to this definition, games are popular media that can communicate issues related to the climate crisis or general environmental concerns.[3] This communication does not require game design specifically catered to environmental concerns, even if such design may improve such communication. Rather, the pervasiveness of the climate crisis requires us to think of all games as inherently caught up in it. This means that ecogames can be categorized as follows:[4]

- Serious games that provide **explicit** education and/or political motivation,
- Artistic works that **implicitly** stimulate ecological reflection,
- Games that do not revolve around, but **incorporate** environmental challenges, or
- Games **representing** elements of ecological crisis.

This diversity of ecogame design direction demonstrates that all games have a part to play in negotiating the present ecological crisis. There is no superiority attached to any of these categories. In fact, ecological consciousness is needed at each of these levels to communicate with the widest range of players possible, but each category entails specific challenges. The ecological communication of serious games is **explicit**. Generally, such games are concerned with what Ian Bogost called *procedural rhetoric*.[5] This term refers to the interactive way in which games can uniquely communicate processes. Games can incorporate player feedback and provide the results on a large virtual scale, while only requiring a slither of the physical world equivalent's time. For this reason, procedural rhetoric has been applied as a strategy for political game design, including ecogame design.[6] According to critics of procedural rhetoric, however, the player is just as much a political figure who makes ethical choices in play, as the developer is while making choices in design.[7] Developers of serious ecogames should likewise be mindful of the possible heterogeneity of their audience. This is especially a concern for ecogame design. Players of ecogames might choose to speed-run to ecosystem collapse or explore if the (ongoing) sacrifice of peoples in the Global South can maintain high living standards in the Global North. Artistic work, when adapted to ecogame design, is generally **implicit** in ecological communication. This means that there is no sense of direct persuasion or education, but rather a sense of players' natural (re)connection to environmental themes. This definition works particularly well for independent games within the genre of "walking simulators," which have historically pushed the boundaries of which types of experiences players consider to be games.[8] When these games replace action with environmental presence that makes players "stop and smell the roses,"[9] these games communicate the value of environments and their preservation to the player. This category could also be packaged in reflective games that do not primarily refer to environmental themes, but do rely on them in their storytelling.[10] One potential pitfall of this form of ecogame design is that it may romanticize nature. If nature is treated as an entity external to us, we risk objectifying it as something of beauty, rather than as implicated in everyday life with agency of its own.[11] Games that do not put ecological crises at the core of the experience may still **incorporate** environmental challenges. Many games that do not claim ecology as their main theme have been read by scholars as actually representing the navigation of challenges of ecological crises. For example, Firaxis Games' *Sid Meier's Civilization VI* added environmental challenges in their *Gathering Storm* expansion pack.[12] Mobius Digital's *Outer Wilds* puts the player to the task of exploring explanations for a solar system's collapse by making the player navigate several planets' ecosystems.[13] These games blend some of the benefits of the previous two categories, namely their direct engagement with ecological crises and non-preachy presentation. However, this also makes their efficacy dependent on the player picking up on these themes. Additionally, they may present the environment as an "issue" to be "fixed" instead of as an integral part of living.[14] These games thus offer widespread opportunities for the implementation of ecogame design, but it is challenging to implement this kind of ecogame design without simplification or trivialization. Finally, games without explicit environmental politics can still **represent** "the fraught socio-environmental conditions of the present."[15] This is possibly the most expansive category, as many games deal with crises that are to some extent manifested in their settings. For example, Hideo Kojima's *Death Stranding* has been studied as featuring a metaphorical connection to our ecological crisis.[16] More provocatively, perhaps, 11 Bit Studios' *Frostpunk* is essentially about the (authoritarian) governance of society during ecological collapse.[17] While such examples may prove fascinating for critics of ecology, their themes may not be evident, let alone considered ecologically relevant, to their average player. The ecological relevance of those games is therefore necessarily accompanied by discussion beyond their play, in social contexts, for example. It could benefit the process of ecogame design to establish the connection to one of these interpretations. Some games, however, exist between, or as a combination of, multiple definitions. For ecogame design, this means potentially more ecological vectors, but also more requisites of balance. In any case, the discussion of these categories should establish that *all games are actually ecological texts*, they all realize their own "mini-ecosystems" with specific game rules, stories, and worlds to explore.[18] If all games are indeed ecological, then all (eco)game design has a responsibility for an attentive consideration. The upside of this point is that, much like Arnaud Fayolle argues in his chapter, one can start incorporating eco-conscious design into any game.

Ecogame Definition 2: Games as Objects of the Environment

Games, like all media, are not just *about* the environment, they are also *of* the environment.[19] Increasingly, games are being studied not because of what is on screen, but what goes on outside of the screen. This "material turn" in approach relates to the remaining two definitions of sustainable games, which are also studied by Thorsten Busch, Florence Chee, and Tanja Sihvonen in their chapters.[20] The second definition of ecogames zooms in on the literal sense of the material, namely games as environmental artifacts. This definition requires the realization of games in an environmentally friendly manner, which includes:[21]

- The neutralization of carbon emissions in **production** (or the development process)
- The minimalization of carbon footprint in the **distribution** of games
- The saving of energy in the **consumption** (or playing) of games
- The limitation and cessation of extraction and waste in the **manufacturing** of games hardware

In addition to their description in these common usage terms, another (ecological) way of thinking about the environmental aspects of games is in terms of life cycle. In this sense, we can distinguish:

- The **pre-life** of games in production and the extraction aspect of manufacture
- The **life** of games in the distribution and consumption of games
- The **afterlife** of games in the waste aspect of manufacture

Finally, there is a distinction to be made between the two essential physical components permeating media—and therefore—games:[22]

- Matter, the tactile substance of space that possesses some degree of mass
- Energy, that which sustains, activates, or otherwise enacts the life of matter.

Having established these ways of looking at the physicality of games, it is now possible to detail how each goal of environmentally friendly game design is relevant. The neutralization of carbon emissions in **production**, or game development, is an especially challenging task as it may compromise ambitions. The global game industry features hundreds of thousands of game development employees and practitioners.[23] In general, games are better if they benefit from interdisciplinary teams. Still, the success of independent (or indie) games demonstrates that even solo developers can find critical and commercial success.[24] The question concerning environmentally friendly game development is not a question of studio size per se, but rather of proportional sustainability. One sample of energy usage is the kWh/capita measure for studios, which should be lower for bigger studios due to possible efficiency measures. In reality, these are typically higher due to their adoption of the latest technological features. Furthermore, other emission figures such as transportation are generally higher among larger studios.[25] Therefore, responsibility of making game development more sustainable lies mostly by leading studios, but should concern every studio. The measures to be taken to neutralize carbon emissions go beyond the scale of this chapter. Still, conceptually speaking, the game development phase is particularly important as its outcomes (i.e., the games) determine most of the other emissions by games. The story of sustainable **distribution** mostly involves the difference between physical and digital distribution of game copies. Findings have diverged on which necessarily uses less energy.[26] On the surface level, the distinction is one between the material and immaterial. However, the Internet relies on server farms which are definitely physical as well, and which, at one point "outperformed" the airline industry in terms of emissions.[27] Physical distribution has its own history of woes. An oft-cited example in media scholarship is the 1983 burial site in New Mexico featuring thousands of unsold Atari game cartridges. The burial site has only been excavated in 2014.[28] More recently, the relative efficiency of digital distribution and the continuing reliance of physical distribution on aerial and naval transportation seem to tip the momentum in favor of increasing digitization.[29] This momentum shift should generally be treated with cautious optimism. Still, it remains necessary to critically assess its universal application, for example considering Internet quality in the Global South.[30] Game **consumption** is a corollary of game design. The act of play cannot be fully controlled by game development. Generally, when play is considered, discourse favors the indirect attribution of sustainability to players' mindsets, as in the textual definition of ecogames.[31] However, the act of play also consumes energy and thus directly engages with sustainability. There are two strategies that could work to improve this situation: visibility and saving. Respectively, they refer to the ability of games to communicate energy usage to the player, and to allow the player to save energy, for example by means of a standby mode or framerate cap.[32] A default "ecomode" can likewise be considered. The implementation of such measures ultimately boils down to design decisions. The pre- and post-life of digital games depends on the **manufacture** of the machines that can run them, including consoles, personal computers, and other machines for which games have been specifically designed. Unfortunately, games hardware relies on rare resources bound to harmful mining practices.[33] The post-life of games hardware is generally one of rubbish: discarded consoles, computers, cartridges, and discs can be found in landfill sites across the world.[34] While it might seem that the manufacture waste of games hardware is a necessary evil in an industry that entertains billions around the world, the scale of the damage is not out of control. The increasingly rapid development of newer consoles with increasingly marginal updates is a direct effect of AAA (triple-A) production of "Bigger and Better" blockbusters.[35] These blockbusters are characterized by improving graphical fidelity, immersive sound, and other new features that may impress the senses. While recent years have changed the conversation of the "console wars" somewhat, with an additional focus on streaming services and alternative (digital) pathways to innovation, the fact remains that new consoles with new and improved features keep influencing the scale of game design.[36] In recent years, there has also been somewhat of a shift in mainstream games production toward subscription models that rely on continuous player patronage.[37] Accordingly, there is an increasing stimulus to release games that will come to their full potential in downloadable expansion packs.[38] Development within and outside of AAA can resist the perpetual upgrade culture of games both by refusing to adhere to new technological standards and business strategies for perpetual revenue and instead opt to refine creative options within current limits. Collectively, this could change the face of not only the game industry but of tech more broadly. Naturally, the considerations of these four subcategories overlap, but, simultaneously, they cannot all be influenced as much by every game's development phase. What remains true is that positive impact can be achieved both directly (cutting down the footprint of development) and indirectly (designing games for less environmentally harmful hardware and energy-saving consumption). Becoming environmentally neutral remains the most urgent and straightforward application of ecogame development.

Ecogame Definition 3: Games as Artifacts of Survivable Industries

Games are not only objects about and of the environment but also artifacts of a social and cultural process. The workforce behind the globe's most popular entertainment form are people, even if the game industry has a habit of not treating them as such. Between massive layoffs of staff, studio closures, and cases of harassment on the work floor, surviving the industry has become increasingly difficult. Games scholarship increasingly frames this precarious situation as a negotiation of sustainability—the ability to sustain work.[39] This is the third definition of sustainable games, and it involves:

- The preservation of **work** in game development
- The **safety** of working as a game developer
- The ability of **survival** as a studio
- The possibility to pursue the individual and/or team's **creativity.**

These definitions are recognizable to many, and possibly not exhaustive, but they may summarize the current drawbacks of the game industry as a work- and marketplace. The preservation of **work** refers to both job preservation and the ability to conduct one's work in comfortable circumstances. Games are computational media. This means they inherently rely on processes of automation. Yet, they are also creative artifacts requiring artistry in terms of writing, audiovisuality, and gameplay design. The aesthetics of games are necessarily balanced between and profiting from these various realizations of creativity.[40] The recent advances of procedural content generation and generative AI (GenAI) as instruments to automate the workforce appear to threaten this balance, tipping the artistry of game development clearly in favor of programmers designing for automation.[41] This is especially concerning as writing and audiovisual design jobs are disproportionately often occupied by women and non-white people, who are still underrepresented in the overall industry. The question of work in game development necessarily means appreciating not only the existence of jobs but also the diversity of their employees. At least as important as being able to work is the **safety** of said work. The game industry has had its fair share of harassment scandals, reflecting a deeply misogynistic game industry.[42] Another facet of unsafe work is "crunch," the condition of significant (unpaid) overtime working for the meeting of a development deadline while maintaining quality standards. While some distinguish between "good" and "bad" crunch, recent scholarship finds all crunch necessarily detrimental to the sustainability of game development work.[43] The safer and healthier the workforce, the more productive and creative the development process, so it is actually a measure of efficiency to not restrain the development team in any such way. The ability of **survival** as a studio is faced by all developers, but some are more precarious than others. "Independent" game developers, with small teams of sizes varying from one sole developer to teams of a couple dozen, find their existence increasingly unsustainable.[44] There are newsfeeds dedicated solely to studio shutdowns.[45] Reasons for the shutdown of studios vary, including acquisition by other studios, commercial failure, or discontinuation because of creative differences or other struggles. Bigger studios may likewise indicate a struggle to meet financial goals. However, when they claim that layoffs are "necessary" to meet these goals, usually this is not an existential necessity but rather a preference in organizational structure to the benefit of executives and shareholders.[46] Studio survivability is therefore in some cases a precarious affair and in others a business strategy. This problem is largely attributable to the structure of platform capitalism.[47] Finally, the **creativity** of game developers partly depends on the state of the industry. The tools employed in game development, for example, the Unity Engine, are actively shaping the kinds of games that can be made.[48] These infrastructures are thus at the same time liberating and restraining independent game production. Small indie games have flooded the market since the mainstreaming of Unity.[49] However, indie games are also increasingly co-opting models of mainstream game development, leading some to term them as "mindie" (mainstream-indie) or "AA" games.[50] These developments show that creativity is constrained in the current modes of production, and game design means walking a tightrope between financial viability and creative expression. In the game industry as a workplace, sustainability means survivability, and this includes being able to work, working safely, staying alive as a studio, and not having to significantly compromise creativity. These conditions are all crucial, but most developers will experience their effects without having much control over their causes. Efforts could involve strengthening the community of game industry workers, for example through unionization.[51] Crucially, any consideration of ecogames must account for how the game industry is about the lives of people whose collective state of being determines the industry's outputs. The importance of building supportive communities for working together is also stressed in the chapters from Aric McBay and eileen mary holowka, and Hexe Fey.

PRACTICAL CONSIDERATIONS FOR THE THREE ECOLOGIES OF GAME DESIGN

While the three definitions of sustainable game design should allow developers to map their ecogame accordingly, this chapter closes with three lessons from academic study with practical implications for ecogame design:

1. Resist the power fantasy
2. Bring **local ecologies** in touch with **planetary realities**—there is no apolitical game
3. **Sustainability** is not easy, there is **no quick fix.**

These concerns likewise demonstrate how the established definitions of sustainable game design come together in the conceptualization of games. The **power fantasy** is a dominant game design paradigm that revolves around giving the player increasing power as the game goes on. There are three critical ways to understand power fantasy: first, it is a fantasy that corresponds to masculine desires of control over life (including over women and the environment), emphasizing individuality instead of co-dependence.[52] Second, as it is a fantasy, it takes the place of real action.[53] Finally, considering the manufacture of games, the power fantasy is also a fantasy of hardware power to increase the representational quality of games.[54] Yet, the closer games try to get to realistic representation of the physical world, the more they consume of those real environments in the process. Power fantasies, following these definitions, reflect development practices of masculine domination, detachment from real engagement, and growing environmental impact. While the valuable craft of progression systems in games must be acknowledged, the task for game developers is to think about these design principles beyond fantasies of power. Consider instead "empowering realities," game design with more breathing room for the dynamics of non-player factors, such as changing environments.[55] The best ecogames are **local** products with a **global** audience. House House's *Untitled Goose Game* was developed by a team of high school friends that resisted urges to expand their studio size after their game became a global media phenomenon.[56] Textually, the game was successful in having players reflect on human-animal relations.[57] Environmentally, the game used limited computational requirements and its physical edition was sustainably packaged.[58] *Untitled Goose Game* may be considered as a successful ecogame spanning the three definitions of sustainable games. Naturally, not all game development contexts have the benefit of a locally cultivated production culture, and some have the disadvantage of no direct global connections. There is no easy solution for this. Game development can be mindful, however, of the necessity of local representatives in the development process by stressing the local contexts of climate change realities. Likewise, this means that in times of climate crisis, there is no more luxury for "apolitical" game design that seeks to distance itself from political realities. Apoliticality is a political stance of dismissal—even the slightest line of dialogue may identify the ecological concerns that audiences rightfully feel.[59] Make sure the content of games is at least in some way relatable to the present times of the climate crisis. Finally, **sustainability** is not easy and requires a continuing sense of responsibility. Some will claim that new innovations will come to the rescue, but **no quick fix** can achieve a sustainable society for all.[60] To illustrate this, the 2025 GDC state of the industry survey indicates that 52% of game companies and 36% of developers use GenAI, whereas 30% of game developers view the technology negatively.[61] GenAI is proclaimed by some as an aesthetic revolution (even for ecogame design!), but this revolution opposes sustainability.[62] Across the three definitions of sustainable game design, GenAI is a destructive force: it destabilizes the industry as a workplace, costs massive amounts of energy and water to function, and is a threat to artistic integrity and local representation with its unreliable scope.[63] In the accumulation of these concerns, it is arguable that *the quick fix of GenAI is at odds with the ecogame on every conceivable level*. Note that these advices are absolutes, whereas, in terms of short-term applicability, progress is more important than perfection. With these final concerns in mind, this chapter hopes to contribute to developers' mapping of how their games correspond to the various definitions of sustainable game design, and that they continue to negotiate climate crisis issues, especially as the crisis intensifies.

NOTES

1 IPCC, "Climate change widespread, rapid, and intensifying – IPCC," *IPCC*, August 9, 2021, https://www.ipcc.ch/2021/08/09/ar6-wg1-20210809-pr/.

2 William J. Ripple, Christopher Wolf, Jillian W. Gregg, Johan Rockström, Michael E. Mann, Naomi Oreskes, Timothy M. Lenton, Stefan Rahmstorf, Thomas M. Newsome, Chi Xu, Jens-Christian Svenning, Cássio Cardoso Pereira, Beverly E. Law, and Thomas W. Crowther, "The 2024 state of the climate report: Perilous times on planet Earth," *BioScience* 74, no. 12 (2024): 812–824, https://doi.org/10.1093/biosci/biae087.

3 Sean Cubitt, *EcoMedia* (Rodopi, 2005).

4 This categorization is slightly modified from: Laura op de Beke, Joost Raessens, and Stefan Werning, "Ecogames: An Introduction," in *Ecogames: Playful Perspectives on the Climate Crisis*, eds. Laura op de Beke, Joost Raessens, Stefan Werning, and Gerald Farca (Amsterdam University Press, 2024), 9–10.

5 Ian Bogost, *Persuasive Games: The Expressive Power of Videogames* (MIT Press, 2007).

6 Clayton Whittle, Trevin York, Paula Angela Escuadra, Grant Shonkwiler, Hugo Bille, Arnaud Fayolle, Benn McGregor, Shayne Hayes, Felix Knight, Andrew Wills, Alenda Chang, and Daniel Fernández Galeote, *The Environmental Game Design Playbook (Presented by the IGDA Climate Special Interest Group)* (International Game Developers Association, 2022).

7 Miguel Sicart, "Against Procedurality," *Game Studies* 11, no. 3 (2011). https://gamestudies.org/1103/articles/sicart_ap.

8 Mia Consalvo and Christopher A. Paul, *Real Games: What's Legitimate and What's Not in Contemporary Videogames* (MIT Press, 2019), 109–130.

9 Jesper Juul, *Handmade Pixels: Independent Video Games and the Quest for Authenticity* (MIT Press, 2019), 201.

10 Kara Stone, "The Earth's Prognosis: Doom and Transformation in Game Design," in *Ecogames: Playful Perspectives on the Climate Crisis*, ed. Laura op de Beke, Joost Raessens, Stefan Werning, and Gerald Farca (Amsterdam University Press, 2024).

11 Timothy Morton, *Ecology without Nature: Rethinking Environmental Aesthetics* (Harvard University Press, 2007).

12 Noam Obermeister and Elliot Honeybun-Arnolda, "Civilization VI: Gathering Storm shows video games can make us think seriously about climate change," *The Conversation*, February 12, 2019, https://theconversation.com/civilization-vi-gathering-storm-shows-video-games-can-make-us-think-seriously-about-climate-change-111791.

13 Lauren Woolbright, "There Is No Planet B: A Milieu-Specific Analysis of *Outer Wilds'* Unstable Spaces," in *Ecogames: Playful Perspectives on the Climate Crisis*, ed. Laura op de Beke, Joost Raessens, Stefan Werning, and Gerald Farca (Amsterdam University Press, 2024).

14 Eric Katz, "The Call of the Wild: The Struggle against Domination and the Technological Fix of Nature," *Environmental Ethics* 14, no. 3 (1992): 265–273, https://doi.org/10.5840/enviroethics199214321.

15 Op de Beke et al., "Ecogames: An Introduction," 10.

16 Víctor Navarro-Remesal and Mateo Terrasa Torres, "Healing a Life out of Balance: Slowness and Ecosophy in *Death Stranding*," in *Ecogames: Playful Perspectives on the Climate Crisis*, ed. Laura op de Beke, Joost Raessens, Stefan Werning, and Gerald Farca (Amsterdam University Press, 2024).

17 Megan Condis and Ben Alfonsin, "*Frostpunk*, the Apocalypse, and the 'Enduring Temptation' of Ecofascism," in *End-Game: Apocalyptic Video Games, Contemporary Society, and Digital Culture*, ed. Lorenzo DiTomasso, James Crossley, Alastair Lockhart, and Rachel Wagner (De Gruyter, 2024).

18 Alenda Y. Chang, *Playing Nature: Ecology in Video Games* (University of Minnesota Press, 2019), 19.

19 Jussi Parikka, *A Geology of Media* (University of Minnesota Press, 2015).

20 Thomas H. Apperley and Darshana Jayemanne, "Game Studies' Material Turn," *Westminster Papers in Communication and Culture* 9, no. 1 (2012): 5–25, https://doi.org/10.16997/wpcc.145.

21 Benjamin J. Abraham, *Digital Games After Climate Change* (Palgrave Macmillan, 2022).

22 Sean Cubitt, *Finite Media* (Duke University Press, 2017).

23 Aphra Kerr, *Global Games: Production, Circulation and Policy in the Networked Era* (Routledge, 2017), 100.

24 Felan Parker, "An Art World for Artgames," *Loading... The Journal of the Canadian Game Studies Association* 7, no. 11 (2013): 41–60, https://journals.sfu.ca/loading/index.php/loading/article/view/119.

25 Abraham, *Digital Games After Climate Change*, 95–113.

26 Kieren Mayers, Jonathan Koomey, Rebecca Hall, Maria Bauer, Chris France, and Amanda Webb, "The Carbon Footprint of Games Distribution," *Journal of Industrial Ecology* 19, no. 3 (2014): 402–415, https://doi.org/10.1111/jiec.12181.

27 Stephen Rust, Salma Monani, and Sean Cubitt, "Introduction: ecologies of media," in *Ecomedia: Key Issues*, ed. Stephen Rust, Salma Monani, and Sean Cubitt (Routledge, 2016), 3.

28 Andrew Reinhard, "Excavating Atari: Where the Media was the Archaeology," *Journal of Contemporary Archaeology* 2, no. 1 (2015): 86–93, https://doi.org/10.1558/jca.v2i1.27108.

29 Joshua Aslan, "Climate Change Implications of Gaming Products and Services" (PhD diss., University of Surrey, 2020, https://doi.org/10.15126/thesis.00853729; Abraham, *Digital Games After Climate Change*, 144–145.

30 David ten Cate, "Review: *Digital Games After Climate Change*, by Benjamin J. Abraham. 2022. Palgrave Macmillan. xv + 254 pp," *Press Start* 9, no. 1 (2023): 98, https://press-start.gla.ac.uk/index.php/press-start/article/view/284.

31 Joost Raessens, "Ecogames: playing to save the planet," in *Cultural Sustainability: Perspectives from the Humanities and Social Sciences*, ed. Torsten Meireis and Gabriele Rippl (Routledge, 2019).

32 Abraham, *Digital Games After Climate Change*, 166–167.

33 Abraham, *Digital Games After Climate Change*, 179–231.

34 Jennifer Gabrys, *Digital Rubbish: A Natural History of Electronics* (The University of Michigan Press, 2013).

35 David B. Nieborg, "Triple-A: The Political Economy of the Blockbuster Video Game" (PhD diss., University of Amsterdam, 2011).

36 Jacob Mertens, "Gaming the System: Digital Revisionism and the Video Game Console Industry" (PhD diss., University of Wisconsin-Maison, 2022).

37 Alexander Bernevega and Alex Gekker, "The Industry of Landlords: Exploring the Assetization of the Triple-A Game," *Games and Culture* 17, no. 1 (2022): 47–69, https://doi.org/10.1177/15554120211014151.

38 Mertens, "Gaming the System," 118–121.

39 Brendan Keogh, *The Videogame Industry Does Not Exist: Why We Should Think Beyond Commercial Game Production* (MIT Press, 2023).

40 Grant Tavinor, *The Art of Videogames* (Wiley-Blackwell, 2009).

41 Aleena Chia, "The artist and the automaton in digital game production," *Convergence: The International Journal of Research into New Media Technologies* 28, no. 2 (2022): 389–412, https://doi.org/10.1177/13548565221076434.

42 Lauren Cho, "The Downward Spiral of the Misogynistic Video Game Industry: It's Truly Up to the 'Last Of Us'," *Loyola of Los Angeles Entertainment Law Review* 42, no. 3 (2022): 175–219, https://digitalcommons.lmu.edu/elr/vol42/iss3/1/.

43 Amanda C. Cote and Brandon C. Harris, "The cruel optimism of 'good crunch': How game industry discourses perpetuate unsustainable labor practices." *New Media & Society* 25, no. 3 (2021): 609–627, https://doi.org/10.1177/14614448211014213.

44 John Banks and Brendan Keogh, "More Than One Flop from Bankruptcy: Rethinking Sustainable Independent Game Development," in *Game Production Studies*, ed. Olli Sotamaa and Jan Švelch (Amsterdam University Press, 2021).

45 "studio closures," GamesIndustry.biz, accessed December 20, 2024, https://www.gamesindustry.biz/topics/studio-closures.

46 "What the current wave of layoffs means for the games industry." GamesIndustry.biz, accessed December 20, 2024, https://www.gamesindustry.biz/what-the-current-wave-of-layoffs-means-for-the-games-industry.

47 Nick Srnicek, *Platform Capitalism* (Polity, 2017).

48 Benjamin Nicoll and Brendan Keogh, *The Unity Game Engine and the Circuits of Cultural Software* (Palgrave Macmillan, 2019).

49 Nadav Lipkin, "Playing with risk: Political-economy, independent games, and the precarity of development in crowded commercial markets," in *Independent Videogames: Cultures, Networks, Techniques and Politics*, ed. Paolo Ruffino (Routledge, 2021).

50 Felan Parker, "Boutique indie: Annapurna interactive and contemporary independent game development," in *Independent Videogames: Cultures, Networks, Techniques and Politics*, ed. Paolo Ruffino (Routledge, 2021).

51 Jamie Woodcock, "Game workers unite: unionization among independent developers," in *Independent Videogames: Cultures, Networks, Techniques and Politics*, ed. Paolo Ruffino (Routledge, 2021).

52 Chad Sean Habel, "Doom Guy Comes of Age: Mediating Masculinities in Power Fantasy Video Games," *M/C Journal* 21, no. 2 (2018): https://doi.org/10.5204/mcj.1383.

53 Jon Bailes, *Ideology and the Virtual City: Videogames, Power Fantasies and Neoliberalism* (Zero Books, 2019).

54 James Newman, *Best Before: Videogames, Supersession, and Obsolescence* (Routledge, 2012).

55 See, for example: Pablo Fraile-Jurado, Marc Llovet-Ferrer, and Fransesc Xavier Roig-Munar, "Toward Realism: An Analysis of Coastal Environments in Open-World Video Games," *Simulation & Gaming* 56, no. 1 (2024): 33–62, https://doi.org/10.1177/10468781241287900.

56 House House, *Untitled Goose Game* [multi-platform] (Panic, 2019); Banks and Keogh, "More Than One Flop from Bankruptcy," 168–169; Brendan Keogh and Rowan Tulloch, "The videogame industry," in *The Media and Communications in Australia*, ed. Bridget Griffen-Foley and Sue Turnbull (Routledge, 2024).

57 Marco Caracciolo, "Animal Mayhem Games and Nonhuman-Oriented Thinking," *Game Studies* 21, no. 1 (2021), https://gamestudies.org/2101/articles/caracciolo.

58 Abraham, *Digital Games After Climate Change*, 84.

59 See also: Mark Bould, *The Anthropocene Unconscious: Climate Catastrophe Culture* (Verso, 2021).

60 Donna Haraway, *Staying with the Trouble: Making Kin in the Chthulucene* (Duke University Press, 2016), 3.

61 GDC, *2025 State of the Game Industry*, 14–16, https://reg.gdconf.com/state-of-game-industry-2025.

62 Stefan Werning, "Generative AI and the Technological Imaginary of Game Design," In *Creative Tools and the Softwarization of Cultural Production*, eds Frédérik Lesage and Michael Terren (Palgrave Macmillan, 2024), 67–90.
63 Kate Crawford, "Generative AI's environmental costs are soaring – and mostly secret," *Nature* 626 (2024): 693, https://doi.org/10.1038/d41586-024-00478-x.

REFERENCES

Abraham, Benjamin J. *Digital Games after Climate Change*. Palgrave Macmillan, 2022.
Apperley. Thomas H., and Darshana Jayemanne. "Game Studies' Material Turn." *Westminster Papers in Communication and Culture* 9, no. 1 (2012): 5–25. https://doi.org/10.16997/wpcc.145.
Aslan, Joshua. "Climate Change Implications of Gaming Products and Services." PhD diss., University of Surrey, 2020. https://doi.org/10.15126/thesis.00853729.
Bailes, Jon. *Ideology and the Virtual City: Videogames, Power Fantasies and Neoliberalism*. Zero Books, 2019.
Banks, John, and Brendan Keogh. "More Than One Flop from Bankruptcy: Rethinking Sustainable Independent Game Development." In *Game Production Studies,* edited by Olli Sotamaa and Jan Švelch. Amsterdam University Press, 2021.
Bernevega, Alexander, and Alex Gekker. "The Industry of Landlords: Exploring the Assetization of the Triple-A Game." *Games and Culture* 17, no. 1 (2022): 47–69. https://doi.org/10.1177/15554120211014151.
Bogost, Ian. *Persuasive Games: The Expressive Power of Videogames*. MIT Press, 2007.
Bould, Mark. *The Anthropocene Unconscious: Climate Catastrophe Culture*. Verso, 2021.
Caracciolo, Marco. "Animal Mayhem Games and Nonhuman-Oriented Thinking." *Game Studies* 21, no. 1 (2021) https://gamestudies.org/2101/articles/caracciolo.
Chang, Alenda Y. *Playing Nature: Ecology in Video Games*. University of Minnesota Press, 2019.
Chia, Aleena. "The Artist and the Automaton in Digital Game Production." *Convergence: The International Journal of Research into New Media Technologies* 28, no. 2 (2022): 389–412. https://doi.org/10.1177/13548565221076434.
Cho, Lauren. "The Downward Spiral of the Misogynistic Video Game Industry: It's Truly Up to the 'Last Of Us'." *Loyola of Los Angeles Entertainment Law Review* 42, no. 3 (2022): 175–219. https://digitalcommons.lmu.edu/elr/vol42/iss3/1/.
Condis, Megan, and Ben Alfonsin. "*Frostpunk*, the Apocalypse, and the 'Enduring Temptation' of Ecofascism." In *End-Game: Apocalyptic Video Games, Contemporary Society, and Digital Culture*, edited by Lorenzo DiTomasso, James Crossley, Alastair Lockhart, and Rachel Wagner. De Gruyter, 2024.
Consalvo, Mia, and Christopher A. Paul. *Real Games: What's Legitimate and What's Not in Contemporary Videogames*. MIT Press, 2019.
Cote, Amanda C., and Brandon C. Harris. "The Cruel Optimism of 'Good Crunch': How Game Industry Discourses Perpetuate Unsustainable Labor Practices." *New Media & Society* 25, no. 3 (2021): 609–627. https://doi.org/10.1177/14614448211014213.
Crawford, Kate. "Generative AI's Environmental Costs are Soaring – and Mostly Secret." *Nature* 626 (2024): 693. https://doi.org/10.1038/d41586-024-00478-x.
Cubitt, Sean. *EcoMedia*. Rodopi, 2005.
Cubitt, Sean. *Finite Media*. Duke University Press, 2017.
Fraile-Jurado, Pablo, Marc Llovet-Ferrer, and Fransesc Xavier Roig-Munar. "Toward Realism: An Analysis of Coastal Environments in Open-World Video Games." *Simulation & Gaming* OnlineFirst (2024). https://doi.org/10.1177/10468781241287900.

Gabrys, Jennifer. *Digital Rubbish: A Natural History of Electronics*. The University of Michigan Press, 2013.

GamesIndustry.biz. "studio closures." Accessed December 20, 2024. https://www.gamesindustry.biz/topics/studio-closures.

GamesIndustry.biz. "What the current wave of layoffs means for the games industry." Accessed December 20, 2024. https://www.gamesindustry.biz/what-the-current-wave-of-layoffs-means-for-the-games-industry.

GDC. *2025 State of the Game Industry*. https://reg.gdconf.com/state-of-game-industry-2025.

Habel, Chad Sean. "Doom Guy Comes of Age: Mediating Masculinities in Power Fantasy Video Games." *M/C Journal* 21, no. 2 (2018). https://doi.org/10.5204/mcj.1383.

Haraway, Donna. *Staying with the Trouble: Making Kin in the Chthulucene*. Duke University Press, 2016.

House House. *Untitled Goose Game*. [multi-platform]. Panic, 2019.

IPCC. "Climate change widespread, rapid, and intensifying – IPCC." *IPCC*, August 9, 2021. https://www.ipcc.ch/2021/08/09/ar6-wg1-20210809-pr/.

Juul, Jesper. *Handmade Pixels: Independent Video Games and the Quest for Authenticity*. MIT Press, 2019.

Katz, Eric. "The Call of the Wild: The Struggle against Domination and the Technological Fix of Nature." *Environmental Ethics* 14, no. 3 (1992): 265–273. https://doi.org/10.5840/enviroethics199214321.

Keogh, Brendan. *The Videogame Industry does not Exist: Why we Should Think beyond Commercial Game Production*. MIT Press, 2023.

Keogh, Brendan, and Rowan Tulloch. "The Videogame Industry." In *The Media and Communications in Australia*, edited by Bridget Griffen-Foley and Sue Turnbull. Routledge, 2024.

Kerr, Aphra. *Global Games: Production, Circulation and Policy in the Networked Era*. Routledge, 2017.

Lipkin, Nadav. "Playing with Risk: Political-Economy, Independent Games, and the Precarity of Development in Crowded Commercial Markets." In *Independent Videogames: Cultures, Networks, Techniques and Politics,* edited by Paolo Ruffino. Routledge, 2021.

Mayers, Kieren, Jonathan Koomey, Rebecca Hall, Maria Bauer, Chris France, and Amanda Webb. "The Carbon Footprint of Games Distribution." *Journal of Industrial Ecology* 19, no. 3 (2014): 402–415. https://doi.org/10.1111/jiec.12181.

Mertens, Jacob. "Gaming the System: Digital Revisionism and the Video Game Console Industry." PhD diss., University of Wisconsin-Maison, 2022.

Morton, Timothy. *Ecology without Nature: Rethinking Environmental Aesthetics*. Harvard University Press, 2007.

Navarro-Remesal, Víctor, and Mateo Terrasa Torres. "Healing a Life Out of Balance: Slowness and Ecosophy in Death Stranding." In *Ecogames: Playful Perspectives on the Climate Crisis,* edited by Laura op de Beke, Joost Raessens, Stefan Werning, and Gerald Farca. Amsterdam University Press, 2024.

Newman, James. *Best Before: Videogames, Supersession, and Obsolescence*. Routledge, 2012.

Nicoll, Benjamin, and Brendan Keogh. *The Unity Game Engine and the Circuits of Cultural Software*. Palgrave Macmillan, 2019.

Nieborg, David B. "Triple-A: The Political Economy of the Blockbuster Video Game. PhD diss., University of Amsterdam, 2011.

Obermeister, Noam, and Elliot Honeybun-Arnolda. "Civilization VI: Gathering Storm shows video games can make us think seriously about climate change." *The Conversation*, February 12, 2019. https://theconversation.com/civilization-vi-gathering-storm-shows-video-games-can-make-us-think-seriously-about-climate-change-111791.

Op de Beke, Laura, Joost Raessens, and Stefan Werning. "Ecogames: An Introduction." In *Ecogames: Playful Perspectives on the Climate Crisis*, edited by Laura op de Beke, Joost Raessens, Stefan Werning, and Gerald Farca. Amsterdam University Press, 2024.

Parikka, Jussi. *A Geology of Media*. University of Minnesota Press, 2015.

Parker, Felan. "An Art World for Artgames." *Loading... The Journal of the Canadian Game Studies Association* 7, no. 11 (2013): 41–60. https://journals.sfu.ca/loading/index.php/loading/article/view/119.

Parker, Felan. "Boutique Indie: Annapurna Interactive and Contemporary Independent Game Development." In *Independent Videogames: Cultures, Networks, Techniques and Politics,* edited by Paolo Ruffino. Routledge, 2021.

Raessens, Joost. "Ecogames: Playing to Save the Planet." In *Cultural Sustainability: Perspectives from the Humanities and Social Sciences,* edited by Torsten Meireis and Gabriele Rippl. Routledge, 2019.

Reinhard, Andrew. "Excavating Atari: Where the Media was the Archaeology." *Journal of Contemporary Archaeology* 2, no. 1 (2015): 86–93. https://doi.org/10.1558/jca.v2i1.27108.

Ripple, William J., Christopher Wolf, Jillian W. Gregg, Johan Rockström, Michael E. Mann, Naomi Oreskes, Timothy M. Lenton, Stefan Rahmstorf, Thomas M. Newsome, Chi Xu, Jens-Christian Svenning, Cássio Cardoso Pereira, Beverly E. Law, and Thomas W. Crowther. "The 2024 State of the Climate Report: Perilous Times on Planet Earth." *BioScience* 74, no. 12 (2024): 812–824. https://doi.org/10.1093/biosci/biae087.

Rust, Stephen, Salma Monani, and Sean Cubitt. "Introduction: Ecologies of Media." In *Ecomedia: Key Issues,* edited by Stephen Rust, Salma Monani, and Sean Cubitt. Routledge, 2016.

Sicart, Miguel. "Against Procedurality." *Game Studies* 11, no. 3 (2011). https://gamestudies.org/1103/articles/sicart_ap.

Srnicek, Nick. *Platform Capitalism*. Polity, 2017.

Stone, Kara. "The Earth's Prognosis: Doom and Transformation in Game Design." In *Ecogames: Playful Perspectives on the Climate Crisis,* edited by Laura op de Beke, Joost Raessens, Stefan Werning, and Gerald Farca. Amsterdam University Press, 2024.

Tavinor, Grant. *The Art of Videogames*. Wiley-Blackwell, 2009.

Ten Cate, David. "Review: *Digital Games After Climate Change*, by Benjamin J. Abraham. 2022. Palgrave Macmillan. xv + 254 pp." *Press Start* 9, no. 1 (2023): 96–100. https://press-start.gla.ac.uk/index.php/press-start/article/view/284.

Werning, Stefan. "Generative AI and the Technological Imaginary of Game Design." In *Creative Tools and the Softwarization of Cultural Production,* edited by Frédérik Lesage and Michael Terren. Palgrave Macmillan, 2024.

Whittle, Clayton, Trevin York, Paula Angela Escuadra, Grant Shonkwiler, Hugo Bille, Arnaud Fayolle, Benn McGregor, Shayne Hayes, Felix Knight, Andrew Wills, Alenda Chang, and Daniel Fernández Galeote. *The Environmental Game Design Playbook (Presented by the IGDA Climate Special Interest Group)*. International Game Developers Association, 2022.

Woodcock, Jamie. "Game Workers Unite: Unionization among Independent Developers." In *Independent Videogames: Cultures, Networks, Techniques and Politics,* edited by Paolo Ruffino. Routledge, 2021.

Woolbright, Lauren. "There Is No Planet B: A Milieu-Specific Analysis of *Outer Wilds*' Unstable Spaces." In *Ecogames: Playful Perspectives on the Climate Crisis,* edited by Laura op de Beke, Joost Raessens, Stefan Werning, and Gerald Farca. Amsterdam University Press, 2024.

Sustainable Game Design in China

14

Pioneering Practices and Future Directions

Vincenzo De Masi, Qinke Di,
Siyi Li, and Yuhan Song

ABSTRACT

This chapter explores the contrast between regulatory and market-driven approaches in fostering sustainable game design within China's growing game ecosystem. As the world's largest gaming market, China faces significant environmental challenges, with data centers consuming 199.07 TWh of electricity in 2020 and projected to reach 490.18 TWh by 2030. With 1.05 billion internet users, the environmental footprint of digital entertainment continues to rise. This chapter compares China's policy-driven framework with Western self-regulation and examines case studies such as Ant Forest, which has planted over 200 million trees. These examples highlight how systematic regulation leads to more measurable outcomes than voluntary corporate actions. As the metaverse market is expected to reach US$12.64 billion by 2027, the need to balance innovation with environmental responsibility becomes more urgent. The Chinese model presents insights that may inform global sustainable development strategies in gaming.

KEY TAKEAWAYS

- **Compare top-down to voluntary regulation.** China's policy-driven approach delivers clearer sustainability results than Western market-based efforts.
- **Efficiency is innovation.** Regulation has pushed Chinese developers to lead in energy-saving technologies for both infrastructure and game design.
- **Sustainability needs more than rules.** Real change requires cultural shifts, international cooperation, and incentives—not just compliance.
- **The future is hybrid.** Blending regulation with creative freedom is the most effective path to sustainable game development worldwide.

INTRODUCTION

China's gaming industry represents a valuable case study for analyzing how regulatory frameworks can shape sustainable innovation in digital entertainment. As of 2023, China remained the world's largest gaming market, generating US$45.2 billion in revenue.[1] This sector, however, brings with it significant environmental challenges. With an internet penetration rate of 73% and more than 1.05 billion netizens,[2] the country's digital infrastructure places considerable pressure on energy consumption and carbon emissions. Data centers supporting gaming activities accounted for 2.7% of China's total electricity use in 2020, and projections suggest this figure will increase to 490.18 TWh by 2030.[3] In contrast, global data centers currently account for roughly 1% of total electricity consumption, highlighting the exceptional scale of China's digital energy footprint.

Unlike Western countries that primarily rely on market incentives and voluntary sustainability efforts by corporations, China has adopted a directive, policy-driven approach. This regulatory model imposes specific environmental and social standards on digital industries, leading to measurable outcomes in several areas. One prominent example is Alibaba's achievement in reducing its Power Usage Effectiveness (PUE) to 1.09 in its Hangzhou data center through immersion cooling techniques, surpassing the global industry average of 1.59.[4] This result reflects the impact of stringent regulatory expectations that go beyond corporate social responsibility declarations.

The emergence of the metaverse in China further complicates the environmental equation. The metaverse is expected to reach a market value of US$12.64 billion by

2027,[5] bringing new demands on computing infrastructure, network systems, and energy consumption. While virtual platforms may reduce emissions from physical events or travel, they also increase the burden on data centers. Balancing these effects requires new sustainability strategies that integrate both regulation and technological optimization.

This chapter explores the tension between China's regulatory model and Western market-driven systems through a comparative analysis supported by case studies. It investigates two opposing approaches to sustainability. The first is a top-down system of policy enforcement that mandates environmental compliance and influences technical design. The second is a bottom-up, market-oriented system where companies pursue sustainability to enhance public image or reduce operational costs. This chapter uses examples such as Ant Forest, which gamifies sustainability and has led to the planting of over 200 million trees.[6]

The goal is to identify which environmental challenges are best addressed through regulatory intervention and which can be solved more effectively through market innovation. The findings offer practical recommendations for developers, infrastructure providers, and policymakers aiming to align digital growth with environmental responsibility.

THE CHINESE GAMING LANDSCAPE AND THE METAVERSE

China's gaming industry represents one of the most relevant examples of how regulatory frameworks can shape environmental outcomes in the digital sector. As mobile gaming accounted for 70% of total industry revenue in 2023,[7] and with the number of mobile gamers expected to reach 993 million in 2024,[8] the scale of digital engagement has major implications for sustainability. This immense user base increases the environmental burden associated with energy consumption, hardware production, and digital infrastructure.

Unlike Western markets, where sustainability practices often rely on voluntary commitments from private firms, China has established a policy-driven model that imposes regulatory obligations. This top-down approach influences how gaming companies operate, leading to the integration of environmental concerns into core business strategies. For example, China's 2021 restrictions on online gaming time for minors,[9] though primarily a social policy, had environmental effects. These restrictions forced developers to build robust monitoring systems and optimize server usage, indirectly contributing to improved energy efficiency.

Chinese companies have responded by adopting specific technical measures. Tencent, one of the country's largest digital platforms, operates the WeChat ecosystem with over 1.3 billion active users.[10] Its game development division, TiMi Studio Group, implemented server consolidation that reduced energy usage by 18% over three years. They also introduced automatic power management systems to lower energy loads during off-peak hours and centralized development tools to avoid resource duplication across projects.[11] Such changes demonstrate how regulatory demands can drive practical improvements beyond superficial corporate sustainability statements.

Similarly, ByteDance has embedded sustainability into its hardware infrastructure. Following its acquisition of VR headset manufacturer Pico,[12] the company committed to achieving operational carbon neutrality by 2030.[13] This includes sourcing 95% renewable energy for its data centers and incorporating energy performance criteria into product development. These decisions reflect an internalization of environmental standards that are absent in many Western contexts.

Alibaba, another key actor, achieved a PUE of 1.09 in its Hangzhou data center by using immersion cooling systems that significantly outperform traditional methods.[14] This efficiency far exceeds the global industry average of 1.59, demonstrating how regulatory pressure can stimulate technical innovation. Immersion cooling reduces energy use by up to 70% compared to conventional air-conditioning systems, showing how infrastructure design aligns with environmental objectives.

Environmental gamification represents another key area of innovation. Ant Forest, developed by Alipay, remains one of the most successful digital sustainability platforms. The system converts low-carbon behaviors, such as walking, using public transit, or reducing electricity use, into points that users can redeem for planting real trees. The program has contributed to over 200 million trees planted across more than 2,900 square kilometers.[15]

Unlike many Western sustainability apps, Ant Forest is deeply integrated into urban infrastructure. Data is automatically collected and verified through partnerships with utility providers, transportation authorities, and payment systems. This allows users to track their individual carbon savings based on verified actions, rather than self-reported behavior. The platform uses visual gamification features such as growing virtual trees, real-world updates, and community sharing, reinforcing pro-environmental habits.[16]

Nonetheless, the regulatory model has also generated unintended outcomes. Faced with increasing compliance costs and limitations on domestic activity, many Chinese gaming firms have expanded internationally. Overseas markets now represent 30% of industry revenue.[17] This shift raises questions about global environmental governance. Without consistent international standards, companies can export environmental burdens to regions with weaker regulations, undermining sustainability goals.

The rise of metaverse technologies in China adds complexity to this context. Unlike the Western model, which often pursues immersive experiences detached from physical infrastructure, the Chinese approach integrates virtual applications with real-world systems. This reflects the goals of the national Smart China strategy,[18] which promotes digital tools to enhance public services and urban planning.

Chinese cities are experimenting with digital twins and augmented reality navigation systems.[19] These applications help optimize urban resources, reduce transportation emissions, and promote sustainable urbanization. For instance, replacing physical meetings with Tencent's virtual platforms or substituting fashion shows with digital equivalents reduces certain emissions.[20] However, these innovations also demand high processing power, shifting energy consumption from the physical to the digital domain. Kshetri and Dwivedi argue that while digital tools can reduce emissions, they may also increase overall energy demand through server activity and network use.[21]

This dynamic presents a challenge for developers. Claims about sustainability must consider the full environmental footprint, including the energy use of data centers, not only the avoided emissions from physical activities. Sustainability assessments should capture both benefits and trade-offs.

ByteDance provides a telling example. While expanding its VR hardware division through Pico,[22] the company simultaneously committed to carbon neutrality. This dual strategy underscores the tension between digital expansion and sustainability. Balancing these goals requires companies to innovate not only in product design but also in infrastructure and verification systems.

Foreign companies have struggled to adapt to this environment. Roblox's failure to meet Chinese regulatory expectations led to withdrawal and staff reductions.[23],[24] These outcomes illustrate the difficulty of operating in a system where sustainability is embedded in policy and infrastructure rather than treated as a peripheral issue.

China's regulatory model has fostered integration between digital systems and environmental planning, with results that surpass many voluntary market efforts. Case studies such as Ant Forest show how directive measures can produce tangible results. At the same time, challenges persist. Effective sustainability requires not only strong regulation but also international cooperation and adaptive innovation strategies.

ENERGY-SAVING TECHNOLOGIES IN GAME DESIGN

Energy-saving technologies are a critical area of focus in the pursuit of sustainable digital entertainment. As gaming infrastructure continues to expand, both in China and globally, energy optimization has become essential. China's approach to this challenge reflects a unique combination of regulatory guidance and technical innovation that contrasts with market-driven models typically found in Western contexts.

One of the most significant developments in this area is the growth of cloud gaming. This model allows users to access high-quality games via streaming rather than requiring powerful local hardware. While this has the potential to reduce energy consumption on individual devices, it increases the demand on centralized data centers. The environmental outcome depends heavily on how efficiently these systems are managed.

China's cloud gaming ecosystem is shaped by the dominance of mobile gaming, which represents 70% of total industry revenue.[25] This creates a more uniform technical environment compared to Western markets, where gaming is distributed across consoles, PCs, and mobile devices. Standardization in China enables developers to optimize platforms with greater precision. Additionally, government-coordinated infrastructure planning ensures that data centers are placed closer to users, reducing transmission energy losses.

Chinese platforms are also subject to specific efficiency requirements. Rather than focusing solely on service speed or visual quality, companies must meet energy performance benchmarks. This regulatory pressure leads to distinctive technical choices. Chinese developers typically prioritize serving large user bases with minimal energy input, whereas Western companies often aim to deliver high-resolution experiences at the cost of greater energy consumption.

Several studies support these observations. Liu et al. showed that cloud gaming,[26] when properly optimized, can reduce system-level energy use, especially when delivering demanding content to low-power devices. Another study by Bangash et al. emphasized the importance of market structure and regulation in achieving these savings.[27] The Chinese model, through coordinated networks and mobile standardization, achieved lower energy consumption per user than its Western counterparts.

Three factors contribute to this outcome. First, unified device ecosystems allow for targeted optimization. Second, infrastructure placement guided by user distribution reduces energy loss. And third, regulatory enforcement encourages higher server utilization, reducing idle energy consumption.

These insights offer practical lessons for global adoption. Western markets could benefit from establishing voluntary efficiency targets and promoting standard protocols for cloud infrastructure. Coordinated planning, even at a regional level, would help optimize energy use without the need for rigid centralization. Furthermore, balancing visual performance with energy efficiency should become a common design goal, not merely a regulatory obligation.

Beyond cloud infrastructure, energy-saving strategies in game development also depend on software optimization. Game engines, the core frameworks for building digital environments, play a critical role. Recent research comparing energy use across engines showed major differences. Unity has a lower consumption than Unreal Engine in the Static Mesh scenario (26%) and in the Physics simulation scenario (351%). However, Unreal Engine has a lower consumption than Unity in the Dynamic Mesh rendering scenario (17%), which means that choosing the appropriate game engine based on specific scenarios and combining their respective advantages is the key to saving energy.[28]

In Western markets, engine optimization is often overlooked due to the economic structure of game production. Developers tend to prioritize rapid development and graphical sophistication, while long-term energy costs are not fully considered. Optimization features such as dynamic memory allocation or asset streaming require more time and expertise, which can increase short-term development costs.

In China, this cost-benefit calculation changes due to regulatory requirements. When energy efficiency becomes a compliance issue, companies invest in optimization early. This shift leads to technical innovation that is later justified by long-term savings in server costs and infrastructure maintenance.

MiHoYo's development of Genshin Impact demonstrates this principle. The company integrated advanced memory management techniques that reduce RAM usage on mobile devices.[29] Although initially more expensive, this optimization reduced server load and lowered operational energy costs. This case shows how environmental regulation can promote technological improvements that are ultimately beneficial from a business perspective.

According to Pérez et al.,[30] global adoption of optimized engines and development practices could reduce energy use by up to 51 TWh per year. This figure highlights the potential scale of impact that software efficiency could have if adopted widely. In China, mobile gaming developers already benefit from lower infrastructure costs due to these efficiency measures.[31]

For Western developers, the lesson is clear. Many optimization practices used in China can be adopted voluntarily. Techniques such as asset compression, intelligent loading systems, and energy-conscious rendering do not require new infrastructure but rather a shift in development priorities. Even without regulatory mandates, these practices offer long-term cost savings and align with growing consumer interest in environmental responsibility.

At the same time, some high-impact changes may not occur without policy incentives. Governments and industry groups in Western countries could introduce sustainability certification programs or offer development subsidies for energy-efficient game production. These measures would encourage companies to adopt sustainable practices without compromising creative freedom.

A combined approach is most effective. Developers should integrate energy-saving strategies into the design process from the beginning, while policymakers support the broader transition through standards and incentives. The Chinese experience illustrates that regulation can act as a catalyst for innovation rather than a barrier.

Energy-saving technologies in cloud infrastructure and software optimization represent a promising direction for sustainable game design. The Chinese experience offers practical insights into what can be achieved through coordinated planning and regulatory innovation. With the right local adaptations, the global gaming industry has the potential to reduce its environmental footprint without compromising creative and commercial goals.

CONCLUSION

This chapter has examined the intersection between regulation, innovation, and sustainability in China's gaming sector. The Chinese model, characterized by strong state-led policies, demonstrates how regulation can shape environmental outcomes in ways that voluntary corporate commitments often fail to achieve. Case studies such as Ant Forest and Alibaba's low-energy data centers illustrate the potential of systemic, measurable impact when sustainability is embedded into infrastructure and user behavior. While China's approach to gaming sustainability demonstrates significant progress across multiple fronts, yet important challenges persist. Issues such as blockchain-related energy consumption, electronic waste, unintended effects of regulation, and shifting market dynamics continue to complicate the path toward a more sustainable industry.

One of the most pressing concerns is the high energy consumption associated with blockchain-based gaming systems. Chinese authorities banned cryptocurrency mining and enforced strict limitations on blockchain use. This decision forced many operators to relocate abroad, often to jurisdictions with greener energy sources.[32] While the domestic environmental impact decreased, the global sustainability implications remain uncertain, raising concerns about displacement rather than the resolution of emissions.

Another major challenge is e-waste, an issue particularly critical in the gaming sector due to frequent device replacement cycles and rapid hardware obsolescence. In

2019, China produced 10.1 million tons of e-waste, making it one of the largest global contributors.[33] Although the country has implemented extended producer responsibility laws requiring manufacturers to meet recycling quotas, the actual recovery rate remains low. Consumer behavior also plays a significant role in shaping the efficacy of recycling systems. Chinese regulations require product labels with disposal instructions and demand that firms provide accessible collection points. However, the frequent desire for the latest devices, driven by aggressive marketing strategies and rapid technological innovation, undermines these efforts. Without a cultural shift toward reuse and repair, even well-designed regulatory frameworks may struggle to achieve impact.

China's regulatory landscape has also introduced practical challenges for game developers. Unlike Western markets, where companies voluntarily engage with environmental goals, Chinese developers must comply with regulatory expectations during the game approval process. This means integrating sustainability considerations into design and infrastructure from the earliest development stages.[34] Although this promotes long-term accountability, it increases initial workload and development costs. Developers often undergo multiple rounds of review, causing delays and resource duplication.

Data localization laws, another pillar of China's digital governance strategy, introduce additional complexity. While designed to enhance cybersecurity and promote domestic infrastructure investment, these policies require foreign companies to duplicate data infrastructure within Chinese territory. This duplication may increase overall energy use and introduce inefficiencies into global cloud systems.[35]

Responding to these obstacles requires not only stronger regulatory frameworks but also deeper systems thinking, well-designed economic incentives, and meaningful cultural shifts. Effective sustainability policy cannot rely solely on top-down enforcement. It must be supported by infrastructure investments, behavioral shifts, and transnational cooperation. China's experience shows the possibilities and limits of directive regulation. Sustainability becomes most effective when it is treated not as a branding tool but as an integrated element of system design and user engagement. The challenge moving forward is ensuring that local success does not result in global displacement of environmental burdens.

Western markets can adopt similar logics without copying China's regulatory system. Structural integration of sustainability, robust verification mechanisms, and full lifecycle assessments can be applied within democratic and decentralized contexts. The key point is that treating sustainability as a core operational concern rather than an add-on increases its effectiveness and resilience. When sustainability is woven into infrastructure, operations, and user engagement, systemic change is possible. Future digital development will depend on structures that prioritize both environmental and technological goals, in order to build ecosystems where environmental outcomes are not side effects, but design imperatives from the start.

Ultimately, a balanced strategy is needed. A hybrid model is likely the most effective path forward. Regulation can set clear sustainability expectations and environmental baselines, while implementation should allow flexibility and creativity so that companies retain freedom to innovate in implementation.

NOTES

1   Newzoo. "Global Games Market Report 2023." Newzoo, February 8, 2024, https://newzoo. com/insights/trend-reports/newzoo-global-games-market-report-2023-free-version.

2   China Internet Network Information Center (CNNIC). "The 52nd Statistical Report on Internet Development in China."

3   Zhou, Feng, Ruimin Wang, and Guoyuan Ma. "Carbon emission scenario analysis of data centers in China under the carbon neutrality target." *International Journal of Refrigeration* 168 (2024): 648–661, https://doi.org/10.1016/j.ijrefrig.2024.09.017.

4   Alibaba Cloud. "Alibaba Cloud Carbon Neutrality Report." Alibaba Group, accessed April 2025. https://sustainability.alibabanews.com/en.

5   Statista. "Metaverse – China." Statista.com, 2025. https://www.statista.com/outlook/amo/ metaverse/china.

6   Wang Chen, "支付宝剑取森林:数字技术如何推动大规模环境保护? [Alipay Ant Forest: How can digital technology promote large-scale environmental protection?]," Dialogue Earth, November 20, 2019. https://dialogue.earth/zh/2/44303/.

7   Giulia Interesse. "China's Gaming Industry: Trends and Regulatory Outlook 2024." China Briefing, February 22, 2024, https://www.china-briefing.com/news/chinas-gaming-industry-trends-and-regulatory-outlook-2024/.

8   Mariam Ahmad, "Fast Plays, Big Pays: Unpacking China's Mobile Gaming Dominance," Global Games Forum, August 9, 2024, https://www.globalgamesforum.com/news/fast-plays-big-pays-unpacking-chinas-mobile-gaming-dominance.

9   State Council of the People's Republic of China. "China tightens measures to prevent online gaming addiction among minors." Chinese government.com, accessed April 25, 2025. https://www.gov.cn/xinwen/2021-08/30/content_5634205.htm.

10 Iris Deng. "Tencent super app WeChat strikes fine balance as everyday tool and public-service platform." South China Morning Post, August 11, 2024, https://www. scmp.com/tech/big-tech/article/3274004/tencent-super-app-wechat-strikes-fine-balance-everyday-tool-and-public-service-platform.

11 Josh Ye. "Tencent said to sharpen focus on metaverse-like developments with advanced new gaming studio." South China Morning Post, October 20, 2021. https://www.scmp. com/tech/big-tech/article/3153035/tencent-said-sharpen-focus-metaverse-developments-advanced-new-gaming.

12 ByteDance. "ByteDance to Acquire Pico, Entering Virtual Reality Field." ByteDance.com, accessed April 25, 2025. https://www.bytedance.com/en/news/bytedance-to-acquire-pico.

13 ByteDance."ByteDanceCommitstoOperationalCarbonNeutralityby2030,"ByteDance.com, accessed April 25, 2025, https://www.bytedance.com/en/news/64101e0917013dda5be8acc6.

14 Alibaba Cloud. "Alibaba Cloud Carbon Neutrality Report." Alibaba Group, accessed April 2025. https://sustainability.alibabanews.com/en.

15 Wang Chen, "支付宝剑取森林:数字技术如何推动大规模环境保护? [Alipay Ant Forest: How can digital technology promote large-scale environmental protection?]," Dialogue Earth, November 20, 2019. https://dialogue.earth/zh/2/44303/

16 Obuobi Bright, et al., "Utilizing Ant Forest technology to foster sustainable behaviors: A novel approach towards environmental conservation," *Journal of Environmental Management* 359 (2024): 121038.

17 Giulia Interesse. "China's Gaming Industry: Trends and Regulatory Outlook 2024." China Briefing, February 22, 2024, https://www.china-briefing.com/news/chinas-gaming-industry-trends-and-regulatory-outlook-2024/.

18 Zhong Yangyang, et al., "The Path to Urban Sustainability: Urban Intelligent Transformation and Green Development—Evidence from 286 Cities in China," *Sustainability* 16, no. 23 (2024): 10394.

19 OpenChat. "Augmented reality and smart cities: Future urbanization (增强现实与智能城市:未来的 urbanization)." Xitu, December 27, 2023. https://juejin.cn/post/7317210704607199283.

20 Laura Husband. "Alibaba's secret to making the metaverse sell." JustStyle, May 10, 2023. https://www.just-style.com/features/alibabas-secret-to-making-the-metaverse-sell/.

21 Kshetri, Nir, and Yogesh K. Dwivedi. "Pollution-reducing and pollution-generating effects of the metaverse." *International Journal of Information Management* 69 (2023): 102620.

22 ByteDance. "ByteDance to Acquire Pico, Entering Virtual Reality Field." ByteDance.com, accessed April 25, 2025. https://www.bytedance.com/en/news/bytedance-to-acquire-pico.

23 Rita Liao. "Two years after pausing service, Roblox China cuts a small number of staff." *TechCrunch, October* 23, 2023. https://techcrunch.com/2023/10/23/roblox-china-tencent-layoff/.

24 Aaron Astle. "Roblox China faces layoffs as the company undergoes evaluation." PocketGamer.biz, October 24, 2023. https://www.pocketgamer.biz/roblox-china-faces-layoffs-as-the-company-undergoes-evaluation/.

25 Giulia Interesse. "China's Gaming Industry: Trends and Regulatory Outlook 2024." China Briefing, February 22, 2024, https://www.china-briefing.com/news/chinas-gaming-industry-trends-and-regulatory-outlook-2024/.

26 Liu Tianyi, et al., "Improving Resource and Energy Efficiency for Cloud 3D through Excessive Rendering Reduction." In *Proceedings of the Nineteenth European Conference on Computer Systems*, pp. 317–332. 2024.

27 Bangash, Ghazal, Pierre-Adrien Forestier, and Loutfouz Zaman. "Cloud Gaming: Revolutionizing the Gaming World for Players and Developers Alike." *Interactions* 31, no. 4 (2024): 54–57.

28 Pérez Carlos, et al., "A comparative analysis of energy consumption between the widespread unreal and unity video game engines," arXiv preprint arXiv:2402.06346 (2024).

29 Kat Bailey. "Why Genshin Impact Dev MiHoYo Is Investing In Fusion Energy." IGN, March 1, 2022, https://www.ign.com/articles/mihoyo-fusion-reactor.

30 Pérez Carlos, et al., "A comparative analysis of energy consumption between the widespread unreal and unity video game engines," arXiv preprint arXiv:2402.06346 (2024).

31 Ma Si. "NetEase leverages gaming prowess to tap into metaverse." China Daily, September 22, 2022. https://www.chinadaily.com.cn/a/202209/22/WS632c0e88a310fd2b29e79399.html.

32 Hedera. "How Much Energy Do NFTs Use?" Hedera.com, accessed April 25, 2025, https://hedera.com/learning/nfts/nfts-energy-use.

33 Forti Vanessa, et al., "The Global E-waste Monitor 2020: Quantities, flows and the circular economy potential." (2020).

34 De La Bruyère, E., D. Strub, and J. Marek. "China'Digital Ambitions A Global Strategy To Supplant The Liberal Order" *National Bureau of Asian Research* (2022).

35 Zhou, Feng, Ruimin Wang, and Guoyuan Ma. "Carbon emission scenario analysis of data centers in China under the carbon neutrality target." *International Journal of Refrigeration* 168 (2024): 648–661, https://doi.org/10.1016/j.ijrefrig.2024.09.017.

SECTION 3

Integrating Environmental Messages into Game Design / Mechanics

Section 3

Design for Environmental Messaging

15

Trevin York

ABSTRACT

This section discusses how games can work toward sustainability with their content
and the way they represent and impact the world.

Stefan Werning opens this section by casting a critical eye at non-critical implementations of sustainability themes in games, demonstrating the risks of harmful misrepresentation and greenwashing. This chapter explores various "antipatterns" that tend
to foster such undesired outcomes, with the aim of helping designers and developers
leverage stronger alternative approaches in their place. This broad overview leads nicely
into the following chapters, which examine specific case studies rooted in the design
and development of games with sustainability messaging and, each in their own way, go
beyond the superficial to integrate sustainability messaging deeply into different aspects
of their design.

The first of these case studies, by Danielle Unéus, explores the integration of sustainability messaging through design in the game *Among Ripples: Shallow Waters*. This
chapter clearly describes the core challenges carried by a traditional genre of gameplay
and the path the development team took to overcome these challenges as they aimed to
craft a "vision of hope."

Next, from Anja Venter, Sam Alfred, and Jonathan Hau-Yoon, comes an in-depth
reflection and examination of the development history of *Terra Nil* centered on the
ways the team's theoretical approach and design philosophy shaped the resulting game.

DOI: 10.1201/9781003512400-18

Throughout, this chapter raises provocations derived from the team's practical experiences aimed at any interested in the practical application of representing sustainability concepts in a digital game.

Third, Clayton Whittle challenges conventional approaches to designing for behavior change with a call to reframe games as platforms that can empower players to act. This chapter illustrates its arguments with original research in collaboration with the game *Subway Surfers*, demonstrating ways in which sustainability messaging in live games can both potentially resonate with or bounce off their existing communities.

Finally, Hanna Wirman opens up this section's topic by drawing together tools from critical and speculative design, art games, and game activism, offering a palette with which developers might approach creating games with complex subject matter. This chapter draws on a wealth of examples from past games, theater, and film, demonstrating how the discussed techniques might be applied across a variety of contexts.

All together, the chapters in this section find that careful, intentional game design for sustainability themes requires a deconstruction and reconstruction of genre conventions and gameplay mechanics. Through these critical examples, we hope you find practically applicable lessons and inspiration from the self-reflective processes described within.

Sustainable Game Design beyond the "Nature-as-Franchise" Paradigm

16

Stefan Werning

ABSTRACT

This chapter explores how the increasing abundance of sustainability-themed games and the sometimes-uncritical application of game design patterns carry a risk of misrepresenting climate emergencies, human-nature relations, as well as alternative, more sustainable futures and might, in the worst case, contribute to greenwashing digital games as a medium. This critical examination is followed by suggested methods for avoiding these risks.

KEY TAKEAWAYS

- A framework for identifying how the overuse of game design patterns and implementing sustainable themes using ecological antipatterns can contribute to greenwashing.
- Approaches to avoid greenwashing and "franchise logic" in sustainable game design, e.g. by strategically juxtaposing more and less critical

DOI: 10.1201/9781003512400-19

gameplay elements or broadening the scope of climate emotions to design for.
- Examples and case studies, with a focus on "cozy" ecogame design, as inspiration on how to identify and overcome "green media franchising."

INTRODUCTION

This chapter explores how the increasing abundance of sustainability-themed games and the sometimes-uncritical application of game design patterns[1] carry a risk of misrepresenting climate emergencies, human-nature relations, as well as alternative, more sustainable futures and might, in the worst case, contribute to greenwashing digital games as a medium.

First, I briefly define why sustainability in games is often handled like a "media franchise" and what that means for how players engage with these games. Then, I show how this approach leads to "antipatterns," i.e., design conventions that look "sustainable" but, e.g., through overuse, can undermine critical thinking. In the final sections, two tactics to address this problem are introduced: these include (1) adapting sustainable design patterns to different game genres, emotional responses as well as constantly changing player expectations and (2) contrasting anti-ecological (but often particularly fun) game design patterns like power fantasies or tech trees with more critical elements, i.e. maximizing players' enjoyment while allowing them to reflect on the potential real-world costs.

In stark contrast to the earlier findings of Abraham and Jayemanne, we are increasingly witnessing an abundance of "climate change games."[2] While AAA video games still rarely fully commit to this topic, AA games and ambitious independent games[3] now commonly thematize ecological sensibilities to reach new audiences and increase their own perceived cultural relevance. Yet, the more "natural" climate themes become for developers to include in their games, the less the incentive to continually reassess and update habitualized design patterns, particularly given often tight production schedules, which may create a numbing sense of familiarity and prevent critical thinking. Meyers and Bittner not only already addressed the potential role of digital games in "teaching young people about sustainability concepts and practices"[4] in 2012 but also already contended that many of these early games like *MiniMonos* and the early MMORPGs *EcoBuddies* and *Pixie Hollow* "contain a veneer of ecological thinking, but rely heavily on [...] commercial, consumerist logic,"[5] which they interpret as "green washing."[6]

Rather than actual greenwashing, which usually implies a deliberate attempt to conceal or confuse, this chapter argues that the non-reflective implementation of sustainability as a theme—even with good intentions—contributes to "green media franchising."[7] The analogous, though not explicitly defined term "pink media franchising" proposed by Derek Johnson,[8] describes how the *Star Wars* media franchise mobilized products like *HerUniverse*, "a line of licensed merchandise launched in 2010 with the aim of carving a feminized space within the franchise" not only to "support[...] a very specific, postfeminist model of consumer lifestyle"[9] but also to "re-inscrib[e] the young female fan in hegemonic gender ideologies."[10] Similarly, from a "green media franchising" perspective, nature and sustainable futures are presented akin to a transmedia franchise, comprising familiar and recognizable imagery, micronarratives, and gameplay such as farming, fishing, and collecting and identifying different types of plants or animals.[11]

While the familiarity of media franchises and the shared "language" they establish can afford critical thinking, franchises more often than not tend to deflect or even co-opt criticism through capitalist mechanisms. For example, Tompkins shows how the *Hunger Games* franchise "cynically mobilize[s] revolutionary desire [and class struggles] as a commercial strategy."[12] A fundamental problem with media franchises, including "green media franchising," is the "reification" of recurring concepts, ideas, and identities. Reification hereby means that, in mainstream discourse and our everyday perception, these ideas lose many of their nuances and increasingly become "a thing," i.e., an immutable unit of meaning directly associated with the medium that monopolized its interpretation. For example, the *Avatar* films and games constitute one of the most widely circulated depictions of (fictional) ecoterrorism in popular culture. This not only (a) makes it easier to draw attention to ecoterrorism as a concept just by evoking the films but also (b) makes it more difficult to draw attention to all the aspects of ecoterrorism that are omitted in this specific depiction.

Similarly, through repeated gameplay, ecologically relevant phenomena like e.g. farming and diving "become reified,"[13] i.e., in mainstream discourse, they become indelibly associated with iconic games like *Stardew Valley* (ConcernedApe 2016) or *Abzû* (Giant Squid 2016). Reification is a powerful framing strategy since the underlying framings are reactivated every time the game is played. However, this also makes it difficult to abstract from and think beyond these increasingly "solidified" interpretations. Even patterns in "environmental game design"[14] can potentially contribute to the reification of ecological concepts if these patterns are applied too routinely or become blunted through overuse.

To explore this challenge, the following section shows how the uncritical implementation of sustainable game design can produce "antipatterns" that foster green media franchising. Identifying these antipatterns can help game makers remain alert and acknowledge where they might slip into uncritical use as well as motivate them to find new design solutions to subvert and contrast overused and, thereby, potentially desensitizing design tropes.

ANTIPATTERNS IN SUSTAINABLE GAME DESIGN AND THE AMBIGUITIES OF COZY ECOGAMES

Previous research on ecogames has highlighted several popular anti-ecological design tropes, including infinitely scaling resource economies in survival-crafting games,[15] frictionless logistics in farming games,[16] or neo-colonial "pleasures" of discovering and conquering unknown territories in strategy games.[17]

These non- or anti-ecological design conventions operate like design patterns, i.e., they are constantly ideated upon and become naturalized through repetition. Moreover, strategically subverting these antipatterns, e.g., by imagining "games representing the environment for their own sake"[18] rather than as a resource, can inspire new forms of sustainable game design. One such example of "placing nature first" by subverting the antipattern of quantifying the utility of in-game environments can be found in this chapter by Anja Venter, Sam Alfred, and Jonathan Hau-Yoon about their design of *Terra Nil* (Free Lives, Clockwork Acorn 2023) in this book.

The term antipattern is not only common in domains like HCI design but has also been used as a frame of reference to identify "dark patterns" in-game design.[19] It refers to "a commonly occurring solution to a problem [...] that looks like a good idea, but backfires badly when applied,"[20] yet—in contrast to dark patterns—without malicious "intent."[21] An increasingly common antipattern that promotes "green media franchise," not least because it has proven itself to be reliably successful on crowdfunding platforms, is the uncritical use of "cozy" game aesthetics in ecogames like *Solarpunk* (Cyberwave 2025), *Loftia* (Qloud Games 2025), or *Outbound* (Square Glade Games 2025). The term "cozy games" is often associated with—or arguably "reified" by—the game *Animal Crossing: New Horizons* (Nintendo 2020) and the role it has played in many people's lives during the COVID-19 pandemic.[22] Apart from a range of distinct visual styles, cozy games feature "no jump scares, timed events, or competitiveness [...] and failure, even if it is present, does not sting."[23] Similarly, narratives usually avoid intense emotional experiences and "even if the story deals with heavier topics or difficult emotions, the gameplay ensures a safe environment to experience them in."[24]

Still, Waszkiewicz and Tymińska argue that cozy games afford "comfort as resistance," in which coziness functions as "an idea of almost radical optimism and unity, which resists irony and opposes a culture dominated by loneliness, cynicism, and posturing."[25] These ideas are rooted in feminist views on "care" and "rest"[26] as political acts. These principles could potentially be useful as an inspiration for ecogame design. Yet, current crowdfunding campaigns for cozy ecogames[27] do not seem to fully deliver on this promise, most notably because they do not put the experience of relaxation and self-care into the context of an ongoing struggle against anti-ecological political and economic systems.

While this chapter primarily discusses "cozy" ecogame design as an example of antipatterns that may foster green media franchising, it is important to also acknowledge the material benefits that cozy games provide. Most visibly, they enable players to mitigate negative climate emotions like "sadness," "grief,"[28] "powerlessness,"[29] or even "shock" and "trauma."[30] In fact, typologies of climate emotions can be a valuable tool for game designers as they demonstrate the spectrum of climate-related feelings or longer term emotional states that games can design for. For example, popular media currently often focus on climate anxiety and negative emotions, but positive climate emotions are similarly varied, ranging, e.g., from "hope" and "joy" to "gratitude," "togetherness," and "care."[31] By systematically exploring this spectrum of climate emotions, also including contrasting emotions to intensify their effect on the player, cozy games can promote resilience and lay the foundation for future action. Moreover, they can provide valuable design inspiration for how to ease new types of players into sustainability as a theme so that future ecogames may draw on the game literacy these new players build up to include more complex and confrontational themes over time.

However, as "cozy" ecogame design often categorically eschews conflict, difficulties, and hardship, it also usually suggests that alternative ways of living, production, and consumption are both possible and reasonably painless. For example, games about living outside of consumerist society like *Among Trees* (FJRD Interactive 2020), *Outbound* (Square Glade Games 2025) or *Camper Van* (Malapata Studio 2025) often downplay risks to personal safety and self-preservation; rather than presenting eating, drinking, and other aspects of "the simulation of the necessary upkeep of maintaining a living body in a somewhat hostile environment" as "mandatory drudgery" like many survival-crafting games do,[32] cozy ecogames often either avoid these aspects altogether or make them negligibly easy. In cozy nature conservation or rewilding games like *Grow: Song of the Evertree* (Prideful Sloth 2021) or *Haven Park* (Bubblebird Studio 2021), nature usually (re)grows fast enough to ensure a smooth, immersive play experience. Cozy games about living in sustainable communities like *Loftia* or *Solarpunk*[33] could potentially foster "ecoliteracy,"[34] i.e. "prepar[e] [the player] to be an effective member of sustainable society." Yet, they reframe and thereby constrain the "Solarpunk" imaginary, a collaborative worldbuilding praxis similarly built on "radical optimism"[35] (497) and practiced across numerous subreddits, fanfiction websites, and other online channels, as a de-facto media franchise, decontextualizing and "reifying" (see above) complex phenomena like wind power, hydroponic farming and upcycling, and often removing the more critical and contentious aspects of Solarpunk storytelling such as political participation and realistic economic alternatives to capitalism.

This is exacerbated by the inclusion of other antipatterns like collectibles and set collection mechanics, either as in-game goals or as achievements (i.e. meta goals). For example, design elements like the Marine Encyclopedia in *Endless Ocean* (Arika 2007) and its sequel,[36] both arguably precursors of the "cozy" approach to environmental game design, are framed as a form of "learning" about marine life. However, since this knowledge is primarily collected, fish are differentiated by rarity and completion of specific categories is tracked, the games present a reductionist view on ecological knowledge. This contrast between nature as a theme and "anti-ecological" mechanics resonates with the argument of Arnaud Fayolle (in this book) that sustainable game design can originate in a theme but needs to be systematically translated into corresponding mechanics. For example, if instead the knowledge were tested and/ or utilized elsewhere in the game, this could make it appear more organic and create intrinsic motivation to learn. In educational games, this approach is often referred to as "intrinsic integration."[37] For example, the first game in the series already includes an aspect of experimentation by requiring players to find out the animals' preferred modes of interaction to unlock additional information and restricts the unlocking of encyclopedia entries to one per day, thereby giving a bit more perceived weight to these learning moments.

Cozy ecogames continually draw audience attention through crowdfunding campaigns and increasing recognizability as subgenres but may also discourage unsuspecting players when real-world sustainable behavior and transitions are not as effortless as they appear in those games. However, systematically subverting cozy game design anti-patterns or juxtaposing them with alternative design sensibilities can potentially be very effective because of the familiarity of "cozy nature" as a de-facto franchise and may help realize the resistive potential of rest and care by incorporating some of the more unpleasant aspects of real life that actually require resistance. One such design idiom could be the more recent inclusion of "dark twist[s]"[38] into the microgenre, which not only retains the pleasures of "cozy" gaming but also creates critical distance and afford a broader range of emotional responses, which according to Nicole Seymour[39] is crucial for climate communication to surprise, subvert expectations, and adapt to quickly changing discursive environments. For example, the fishing game *Dredge* (Black Salt Games 2023) arguably owed its success to the combination of "cozy" fishing mechanics with darker, Lovecraftian themes. Together, these experiences of awe and terror arguably capture the human fascination with the ocean more aptly, and more productively, than the serene seascapes of more traditional cozy ecogames like e.g. *Coral Island* (Stairway Games 2022). Replacing the supernatural horrors with more man-made (climate) threats could be a way to build on this "formula" and attempt to harness the political potential of "cozyness" in games.

While this chapter focuses on commercial games, it should be noted that similar instances of sustainable antipatterns can also be found in modifications ("mods") of popular games. Despite the critical potential of ecomodding,[40] i.e. mods that explore human-nature relationships in ways that the original video games do not afford, even well-intentioned mods may perpetuate ambivalent or even anti-ecological ways of thinking.[41] For example, the mod "One with Nature" for *The Elder Scrolls V: Skyrim* (Bethesda Game Studios 2011),[42] which "is designed to give [players] control over how certain animals and creatures react to [their] character," may be customized to explore complex human-animal relations but—as indicated in the mod's screenshots—is often used to enact romanticized notions of bonding with wild animals, in line with cozy game aesthetics. The mod "Wild Animals Companion" for *Red Dead Redemption 2* (Rockstar Games 2018) fulfills a similar purpose, while other mods for the same game such as "Spawn Legendary Animals & Fish" facilitate the built-in set collection gameplay, thereby removing the element of mystery from the most elusive in-game creatures, which partially defied the otherwise utilitarian representation of animals in the original game. Exploring the impact of "green media franchising" and antipatterns on co-design practices such as modding is outside the scope of this chapter but would constitute a fruitful direction for further research.

ADAPTING DESIGN PATTERNS TO GAME GENRES AND CHANGING PLAYER EXPECTATIONS

To avoid falling back on treating sustainability as a "media franchise," repeating antipatterns or even fostering "greenwashing," it is important to more systematically consider the context in which sustainable game design patterns are applied. For this purpose, design idioms are a useful supplementary concept. In software development, idioms commonly refer to lower level design solutions that are "tied to a particular programming language,"[43] similar to how idioms commonly refer to recurring expressions of an idea in a specific natural language. Adapted to game design, idioms can describe implementations of a pattern that are specific to the design "language" of a given game type, genre, or intellectual property they are applied to.

For example, the sustainable game design pattern of "no-win scenarios"[44] can and should be implemented differently in a strategy game as opposed to a first-person shooter or in a competitive as opposed to a cooperative game. As argued by Deterding,[45] players' interpretations of such idioms also differ considerably based on how the game is framed, advertised, and reviewed; consider, e.g., a very short free-to-play news game like *September 12th* (Gonzalo Frasca 2003) referenced in the Environmental Game Design Playbook compared to a commercial game sold as an entertainment product. Commercial games like *Frostpunk* (11 Bit Studios 2018) usually only implement localized no-win scenarios, e.g. dilemmas in which no "correct" choice is available, but a no-win scenario that renders playing meaningless altogether as in *September 12th*—which would effectively question the agreed-upon status of games as entertainment-oriented consumer products—would likely not be accepted and disrupt the relationship between player, game, and developer. The Playbook implicitly acknowledges this, e.g., by urging developers to avoid "breaking the player-designer trust agreement"[46] by abusing this tactic.

Yet, while the Playbook suggests that games using no-win scenarios should generally "goad the player into thinking, even briefly, that it is possible to beat [the] game," different implementations may evoke other but similarly productive cognitive and emotional responses. For example, clearly communicating the inevitability of loss as in games like *Oiligarchy* (Molleindustria 2008) or board games like *Kyoto* (Sabine Harrer and Johannes Krenner 2020) can afford introspection and critical engagement with concepts like petrotemporalities,[47] for example anticipating and preparing for (or trying to ignore) the impending end of oil culture, specifically because players already expect an unfavorable ending, which triggers a different emotional response. Moreover, compared to their initial application as a form of "transgressive" game design, "no-win" scenarios have gradually become an established design pattern. Without switching up design idioms used to implement them, these scenarios can easily come across as tropes, contribute to negative climate emotions like "feeling betrayed" or "resentment,"[48] and evoke cynicism or indifference.

One example of "playing with" the pattern and retaining its critical potential could be to implement no-win scenarios randomly (but not in the first few attempts at a given scenario) to maintain their effect of perceived "unfairness" that—if applied to real-world economic constellations and the "uphill battle" of climate politics—can become very relatable and impactful. Apart from tailoring sustainable game design patterns to changing contexts, a second approach to avoid green media franchising involves contrasting anti-ecological and more reflective (but less familiar and potentially less appealing) design patterns in the same game, game franchise, or microgenre, as will be elaborated upon in the following section.

BEYOND "GREEN MEDIA FRANCHISING": CONTRASTING MORE AND LESS SUSTAINABLE GAME DESIGN

Above, I have argued that quickly iterating on sustainability as a theme in video games can unintentionally lead to misrepresentation or even reduce human-nature relations and climate futures to a mere "media franchise" that helps sell more copies in the short term. For example, acts of natural conservation and sustainable lifestyles—as presented in many contemporary "cozy" ecogames—can come across as busywork and prevent critical reflection if they are (a) overused in the same game and/or (b) overused in the same microgenre (like farming or slice-of-life games) over time. That may seemingly paradoxically make them more difficult to critically engage with, a phenomenon I had summarized earlier using the concept of reification. In that case, sustainable themes are perceived and "consumed" by players in a similar way as the familiar but often "harmless" stories and characters of fictional franchises like Marvel or Star Wars.

As shown above, even previously disruptive design patterns like no-win scenarios over time become an accepted "convention" of procedural rhetoric, and popular game design frameworks like cozy farming games gradually turn into rigid "microgenres," which limits their critical potential as "green activations."[49] To avoid this, it is important to not only look for positive examples but also deliberately identify environmental "anti-patterns" that emerge from standardized design practices or from incorporating trends like cozy game aesthetics without considering their implications for players' environmental awareness and identity. Strategically subverting these antipatterns—e.g., juxtaposing cozy design tropes with darker plot twists or more confrontational gameplay—may be a way to use these conventions "against" players, create more impactful experiences, and, thereby, ideally generate more controversial discourse in gaming-related social media channels, which would contribute to maximizing these games' outreach.

Moreover, sustainable design patterns and antipatterns should be strategically juxtaposed and balanced rather than being used as criteria to simply distinguish "good" from "bad" ecogames. Anti-ecological gameplay tropes like "automation" and "tech trees,"[50] which perpetuate problematic narratives of techno-optimism and unfettered progress, are often deeply engaging and rewarding, not least because they have been iterated upon and refined for decades. Rather than denying players this enjoyment, by mixing positive and negative design patterns, designers can use the former to critique and/or defamiliarize the latter, thereby creating critical tension. This strategy serves a purpose that Anne-Marie Schleiner has described as the "broken toy tactic," i.e., choosing design approaches that "snap the player out of the hypnotic circle of toy operationality"[51] afforded by many game worlds, which poses a "challenge to critical and activist game design."[52]

For example, the "reverse city builder" *Terra Nil* also exhibits aspects of cozy game design like non-competitive gameplay and calming audiovisual design, but the unspecific setting and lack of a human protagonist or focalization help create critical distance and avoid framing rewilding as a simple and effortless solution that mankind can "apply" any time. Similarly, popular gameplay tropes like "power fantasy"[53] experiences can be juxaposed with other design idioms that highlight their limitations if translated into real life, e.g., by temporarily granting similar unconditional growth to in-game enemies.

More broadly speaking, perspective switching, which draws on and expands upon the design pattern of "conflicting goals,"[54] can be a powerful tool to contrast positive and negative environmental game design. For example, the narrative adventure game *Fahrenheit* (Quantic Dream 2005) uses it effectively, albeit not in an ecological context, by requiring the player to control both a fugitive falsely accused of murder and a cop charged with arresting them. This type of split identification can work both, as in *Fahrenheit*, via consecutive phases—consider e.g. strategy games like the *Civilization* franchise, forcing players to control both a traditional expansionist and an indigenous faction in the same area—or as a cooperative asymmetrical multiplayer game, in which players must live with the consequences of each other's choices. Such contrasting, hybrid design approaches can help prevent sustainable game design from becoming reified and blunted over time and avoid reducing sustainability to a mere media franchise.

To conclude, while it is important to be aware of how overusing sustainable game design patterns can foster "green media franchising," implementing this insight into design practice requires careful balancing as demonstrated by the cozy ecogames chosen as a case study. If they provide a place of rest and relaxation, as well as enable players to cope with exhaustion and to negotiate negative and positive climate emotions, that in itself can have political implications. Representing characters that seemingly resist the need to "work or 'struggle'" in games can be a radical proposition in a societal context shaped by neo-liberal capitalism. Yet, many cozy ecogames utilize gameplay idioms that do perpetuate ideas of commodification, ownership, and progress, e.g. *Stardew Valley*'s systems for upgrading tools or the Shipping Collection tab (and associated in-game achievements) that tracks "productivity" as a function of crops and resources sold. Thus, the challenge for game designers will be to acknowledge the potential for greenwashing that comes with an increasing number of sustainability-themed games and to keep iterating on, balancing, and occasionally "breaking" sustainable game design patterns to enable players to enjoy their games while retaining a critical disposition.

NOTES

1 Whittle, Clayton, Trevin York, Paula Angela Escuadra, Grant Shonkwiler, Hugo Bille, Arnaud Fayolle, Benn McGregor, et al. The Environmental Game Design Playbook (Presented by the IGDA Climate Special Interest Group). International Game Developers Association, April 6, 2022. https://igda.org/resources-archive/environmental-game-design-playbook-presented-by-igda-climate-special-interest-group-alpha-release/.
2 Abraham, Benjamin, and Darshana Jayemanne. "Where Are All the Climate Change Games? Locating Digital Games' Response to Climate Change." *Transformations* 30, no. 1 (2017): 74–94.
3 Sometimes referred to as "III" games; see e.g. https://www.gamesindustry.biz/an-era-of-triple-i-development-is-almost-here.

4 Meyers, Eric M., and Robert Bittner. "'Green Washing' the Digital Playground." In *Proceedings of the 2012 iConference*, 608. ACM, 2012.

5 Meyers and Bittner, 610.

6 Meyers and Bittner, 608.

7 Werning, Stefan. "Die Natur als Transmediales Franchise? Zu den Möglichkeiten und Grenzen von Ecogames." *Montage A/V* 32, no. 2 (2024): 49–70.

8 Johnson, Derek. "'May the Force Be with Katie': Pink Media Franchising and the Postfeminist Politics of HerUniverse." *Feminist Media Studies* 14, no. 6 (2014): 895.

9 Johnson, 896.

10 Johnson, 906.

11 For an overview of recurring themes, see e.g. games filtered by the tag 'nature' on Steam (even though not all of these games exhibit this 'franchise' logic): https://store.steampowered.com/tags/en/Nature/.

12 Tompkins, Joe. "The Makings of a Contradictory Franchise: Revolutionary Melodrama and Cynicism in The Hunger Games." *Journal of Cinema and Media Studies* 58, no. 1 (2018): 70.

13 Lakoff, George. "Why It Matters How We Frame the Environment." *Environmental Communication* 4, no. 1 (2010): 70–81.

14 Whittle et al.

15 Abraham, Benjamin. Digital Games After Climate Change. Palgrave Macmillan, 2022.

16 Werning 2024.

17 Mukherjee, Souvik. "No Cyclones in Age of Empires: Empire, Ecology, and Video Games," in *Ecogames. Playful Perspectives on the Climate Crisis*, eds. Laura op de Beke, Joost Raessens, Stefan Werning, and Gerald Farca (Amsterdam University Press, 2024).

18 Abraham, 61.

19 Zagal, José P., Staffan Björk, and Chris Lewis. 2013. "Dark Patterns in the Design of Games." In Foundations of Digital Games 2013. https://www.diva-portal.org/smash/record.jsf?pid=diva2%3A1043332.

20 Widdicks, Kelly, Daniel Pargman, and Staffan Björk. "Backfiring and Favouring: How Design Processes in HCI Lead to Anti-Patterns and Repentant Designers." In *Proceedings of the 11th Nordic Conference on Human-Computer Interaction: Shaping Experiences, Shaping Society*, 1–12. ACM, 2020. https://doi.org/10.1145/3419249.3420175.

21 Widdicks et al, 3.

22 Waszkiewicz, Agata, and Marta Tymińska. "Cozy Games and Resistance Through Care." *Replay: The Polish Journal of Game Studies* 11, no. 1 (2024): 7–16, https://doi.org/10.18778/2391-8551.11.01.

23 Waszkiewicz and Tymińska, 2024, 8.

24 Waszkiewicz and Tymińska, 2024, 8.

25 Waszkiewicz and Tymińska, 2024, 11.

26 Waszkiewicz and Tymińska, 2024, 12.

27 Successful sustainability-theme "cozy" game campaigns launched in 2024 on Kickstarter alone include Tales of Seikyu—cozy farming game, Camper Van: Make it Home, Outbound - Cozy Camper-Van Exploration-Crafting Game, Loftia—a cozy online game set in a warm, solarpunk world, Aspens: A cozy strategy game about growing a tiny forest and Sugardew Island - Your cozy farm shop.

28 Pihkala, Panu. "Toward a Taxonomy of Climate Emotions." *Frontiers in Climate* 3 (January 2022). https://doi.org/10.3389/fclim.2021.738154.

29 Pihkala, 12.

30 Pihkala, 11.

31 Pihkala, 8.

32 Abraham, Benjamin. Digital Games After Climate Change (Palgrave Macmillan, 2022), 71.

33 See https://kickstarter.com/projects/cyberwave/solarpunk-survival-craft-game-in-world-of-floating-island.
34 McBride, Brooke Baldauf, Carol A. Brewer, A. R. Berkowitz, and William T. Borrie. "Environmental Literacy, Ecological Literacy, Ecoliteracy: What Do We Mean and How Did We Get Here?" *Ecosphere* 4, no. 5 (2013): 1–20.
35 Schuller, W. Kees. "Solarpunk: Radical Optimism as Praxis," in *Routledge Handbook of Climate Change and Society*, ed. Steven Brechin and Seungyun Lee (Routledge, 2024), 497.
36 See https://endlessocean.fandom.com/wiki/Marine_Encyclopedia_(Endless_Ocean).
37 Habgood, M. P. J., and S. E. Ainsworth. "Motivating Children to Learn Effectively: Exploring the Value of Intrinsic Integration in Educational Games." *Journal of the Learning Sciences* 20, no. 2 (2011): 169–206. https://doi.org/10.1080/10508406.2010.508029.
38 See e.g. https://www.inverse.com/gaming/cozy-games-summer-game-fest-dark-twist-wanderstop-grave-seasons.
39 Seymour, Nicole. *Bad Environmentalism: Irony and Irreverence in the Ecological Age* (University of Minnesota Press, 2018).
40 Bohunicky, Kyle Matthew. "Ecomods: An Ecocritical Approach to Game Modification." *Ecozon@: European Journal of Literature, Culture and Environment* 8, no. 2 (2017): 72–87. https://ebuah.uah.es/dspace/handle/10017/31327.
41 Werning, Stefan. "Ecomodding: Understanding and Communicating the Climate Crisis by Co-Creating Commercial Video Games." *Communication +1* 8, no. 1 (2021). https://scholarworks.umass.edu/cpo/vol8/iss1/7/.
42 See https://www.nexusmods.com/skyrimspecialedition/mods/13343.
43 Gaelli, Markus, Oscar Nierstrasz, and Serge Stinckwich. "Idioms for Composing Games with EToys." In Fourth International Conference on Creating, Connecting and Collaborating through Computing (C5'06), 222–31. IEEE, 2006. https://doi.org/10.1109/C5.2006.20.
44 Whittle et al., 35.
45 Deterding, Sebastian. "The Mechanic Is Not the (Whole) Message: Procedural Rhetoric Meets Framing in Train & Playing History 2." *DiGRA/FDG '16 - Abstract Proceedings of the First International Joint Conference of DiGRA and FDG* 13, no. 2 (2016). http://www.digra.org/digital-library/publications/the-mechanic-is-not-the-whole-message-procedural-rhetoric-meets-framing-in-train-playing-history-2/.
46 Whittle et al., 36.
47 Kristoffersen, Berit, Gavin Bridge, and Philip Steinberg. "Time for Oil: Competing Petrotemporalities in Norway's Lofoten/Vesterålen/Senja Archipelago," in *Cold Water Oil*, eds. Fiona Polack and Danine Farquharson (Routledge, 2021), 176–193.
48 Pihkala, 8.
49 The term is defined by the Playing for the Planet alliance as "new features and messaging […] which highlight environmental themes such as conservation and restoration"; see https://www.unep.org/news-and-stories/press-release/2nd-annual-green-game-jam-brings-leading-game-companies-together.
50 See e.g. https://tvtropes.org/pmwiki/pmwiki.php/Main/GameplayAutomation and https://tvtropes.org/pmwiki/pmwiki.php/Main/TechTree respectively.
51 Schleiner, Anne-Marie. "The Broken Toy Tactic: Clockwork Worlds and Activist Games," in *The Playful Citizen: Civic Engagement in a Mediatized Culture*, eds. René Glas, Sybille Lammes, Michiel de Lange, Joost Raessens, and Imar de Vries (Amsterdam University Press, 2019), 125.
52 Schleiner, 130.
53 See e.g. https://tvtropes.org/pmwiki/pmwiki.php/Main/PowerFantasy.
54 Whittle et al., 47.

From Exploitation to Restoration

17

Rethinking Management Games

Danielle Unéus

ABSTRACT

Exploring sustainability through game design, Among Ripples: Shallow Waters, an eco-tycoon simulation by the independent studio Eat Create Sleep, transforms environmental challenges into interactive, restorative experiences. Rather than reinforcing bleak narratives of ecological collapse, the game fosters a sense of agency and optimism, encouraging players to rethink humanity's relationship with ecosystems and sustainability. This chapter examines the design challenges encountered, the solutions developed, and the practical insights that may guide other developers in embedding sustainability into their projects.

KEY TAKEAWAYS

- Actionable insights derived from an eco-tycoon simulation game's development, stemming from lessons learned from the design challenges the team encountered when trying to embed sustainability into their project.

DOI: 10.1201/9781003512400-20

BACKGROUND

Eat Create Sleep is a game studio with expertise in simulation games. In the years leading up to the pandemic, our team, like many others, wrestled with environmental anxiety and the grim narratives surrounding the future of our planet. We envisioned a project that offered hope—an eco-management game that highlighted humanity's capacity for positive action and collaboration. This vision led to the development of *Among Ripples: Shallow Waters*,[1] a game where players become part of the Freshwater Biome Restoration Division, working with scientists to rehabilitate fragile ecosystems. We wanted to emphasize the real-world experts who are tackling these issues today. We made scientists the heroes of our game, placing their knowledge and expertise at the heart of the player's journey. To make the simulation as authentic as possible, we collaborated with ecologists, grounding the game in realistic ecological processes to show that this kind of restoration happens in real life.

Players experiment with the interactions between flora and fauna—restoring biodiversity and water quality in lakes. We were inspired by games such as Zoo Tycoon,[2] Planet Coaster,[3] and Frostpunk.[4] However, as we moved from concept to mechanics, we encountered a central challenge: How do you design a management game that values ecological restoration and sustainability when the genre's traditional mechanics are rooted in unsustainable practices?

THE DESIGN PROBLEM

In an early prototype of *Among Ripples: Shallow Waters*, players managed a lake that initially consisted of just water. The gameplay allowed them to create an environment for fish to thrive by introducing flora, objects, terraforming, and balancing water quality. Each of these actions costs money, with players starting with a set sum and earning more by successfully introducing different fish species. However, this led to an unintended outcome: players were incentivized to optimize their lake for capital gain, often transforming it into a monoculture of dace for maximum profit. The lakes became commodities, optimized for financial efficiency rather than ecological balance.

From a previous release of a game at Eat Create Sleep, we learned a critical lesson: the importance of meeting player expectations. While it may seem obvious, positioning a game as part of a specific genre, like a management game, means the audience will anticipate a certain type of gameplay experience. Straying too far from those expectations can leave players feeling frustrated or even cheated. With *Among Ripples: Shallow Waters*, we took this lesson seriously, conducting extensive research to deeply understand what is core to popular management games and where we could innovate without alienating our player base.

What we realized from this was that almost all management games are inherently built around capitalist economic systems. Players are typically tasked with managing resources like money, materials, and manpower, focusing on maximizing efficiency and profit. Core mechanics—such as building and upgrading, managing supply and demand, and ensuring resource flow—all revolve around maintaining financial growth. In these types of games, much like in capitalism, systems are optimized toward profit, rarely having to worry about diversity or sustainability. It took us some time to understand how deeply ingrained these economic principles are in this genre of game design—a challenge highlighted by Dyer-Witheford and de Peuter[5]—making it particularly difficult to create a system that prioritizes sustainability over profit.

From our perspective, capitalism is one of the key drivers behind the destruction of our ecosystems. By simply adopting the typical economic systems of management games, we unintentionally created a game that trained players to exploit ecosystems for profit, rather than restore them. Where most management games revolve around maximizing profit, our goal became to shift the focus toward maximizing life and biodiversity.

SOLUTION

To make this shift, we overhauled several core mechanics and rethought the role of money in our game. The first crucial step was ensuring that lakes, flora, and fauna could not be used as capital-generating assets.

We moved away from a traditional profit-driven economy, replacing it with a grant-based system where progression is tied to ecological impact rather than financial success. Instead of accumulating wealth, players advance by refining their team's expertise, expanding their restoration efforts, and uncovering the evolving narrative. Each successfully rehabilitated lake grants them the opportunity to take on new, more complex ecosystems, reinforcing the idea that progress is about stewardship rather than ownership. Alongside restoration, players can invest in biodiversity research and develop their team's skills in ecological management, species conservation, water quality analysis, and sustainable restoration techniques.

Another major shift was in the game's setting. Rather than grounding the story in present-day struggles, we chose to imagine a near future where humanity had confronted the environmental crisis as an existential threat and taken decisive action. In that world, the focus is no longer on mere survival but on restoring ecosystems and healing the damage left behind by capitalism. This narrative shift sparked in us a greater creativity in imagining new futures, moving us beyond the typical apocalyptic narratives that dominate discussions about the planet. Instead, we aimed to craft a vision of hope—one where humanity is actively repairing and regenerating the world. *Solar punk*[6] was a key inspiration for this project, offering a framework for envisioning a future where technology, sustainability, and ecological harmony coexist.

Our vision of alternative futures shaped not just our narrative but our entire approach to game design. By stepping away from conventional economic systems, we discovered new ways to engage players. Here are the lessons we learned that extend beyond our game project—insights that other developers could apply to integrate sustainability into their own designs.

PRACTICAL ADVICE FOR DEVELOPERS

- **Recognize that mechanics express values**: Game mechanics shape how players interact with virtual environments, influencing their perception of ecological systems. As Alenda Y. Chang highlights in *Playing with Nature*,[7] games have the power to reinforce or reshape how we engage with nature. Critically examine whether your systems promote unsustainable practices and explore alternatives that align with your environmental values. For a wealth of tools and frameworks on designing with values in mind, *Values at Play*[8] is an excellent resource.
- **Engage experts**: Collaborate with specialists (in our case, ecologists and limnologists) to ground your game in real-world processes, ensuring authenticity and educational value while highlighting the work of those tackling these issues in reality.
- **Balance innovation with player expectations**: Understand your target audience's expectations and identify areas where you can innovate without alienating players. With *Among Ripples*, we emphasized familiar gameplay for onboarding while gradually introducing more innovative mechanics as players progressed.
- **Redefine success metrics**: Shift success metrics away from profit maximization toward more meaningful ecological goals, such as biodiversity, environmental restoration, and sustainability.
- **Emphasize stewardship over ownership**: Design systems that reward players for being caretakers rather than exploiters of resources, reinforcing the idea that progress comes through nurturing environments rather than depleting them.
- **Create hopeful futures**: Challenge apocalyptic environmental narratives by envisioning positive futures where humanity actively repairs and regenerates ecosystems, inspiring players to see possibility rather than inevitability.

Among Ripples: Shallow Waters taught us that sustainability in game design goes beyond surface-level themes—it demands a fundamental shift in genre conventions and core mechanics. By challenging profit-driven systems and redefining success, the game explores new ways to engage players in ecological restoration, sustainability, and stewardship.

More than just a backdrop, sustainability in games can become a powerful, interactive medium for reimagining humanity's connection to nature. By reshaping how players interact with virtual ecosystems, games have the potential to inspire broader conversations and meaningful change in how we approach real-world environmental challenges.

NOTES

1 Eat Create Sleep. *Among Ripples: Shallow Waters*. PC. Forthcoming.
2 Blue Fang Games. *Zoo Tycoon*. Microsoft Studios. PC. 2001.
3 Frontier Developments. *Planet Coaster*. Frontier Developments. PC/Console. 2016.
4 11 bit studios. *Frostpunk*. 11 bit studios. PC/Console. 2018.
5 Dyer-Witheford, Nick, and Greig de Peuter. *Games of Empire: Global Capitalism and Video Games*. Minneapolis: University of Minnesota Press, 2009.
6 Robinson, Kim Stanley. "Solarpunk and the Pedagogical Value of Utopia." *Sustainability: The Journal of Record* 13, no. 2 (2020): 79–81. Accessed March 14, 2025. http://www.susted.com/wordpress/content/solarpunk-the-pedagogical-value-of-utopia_2020_05/.
7 Chang, Alenda Y. *Playing Nature: Ecology in Video Games*. Minneapolis: University of Minnesota Press, 2019.
8 Flanagan, Mary, and Helen Nissenbaum. *Values at Play in Digital Games*. Cambridge: MIT Press, 2014.

Terra Nil

Reclaiming the Wasteland through Affective Play

18

Dr Anja Venter, Sam Alfred, and Jonathan Hau-Yoon

ABSTRACT

City-builder and strategy computer games have become a mainstay of the contemporary gaming landscape—offering an imaginary of planning and governance in made-up worlds, along with the public's participation in such processes. However, from an ecological perspective, these games frequently brush over pertinent issues such as the exploitation of both human and environmental resources, the colonization of land, and the unsustainable culture of unfettered expansion. This chapter explores the development of South African indie game Terra Nil, as an antidote to these genre's blind spots, centered around the question: "What if we put nature first?" This chapter provides an overview of the development history of the game by examining how player feedback shaped new ideas in a familiar genre, using emotional impact as a guiding force for mechanics, art, and sound—affecting player feelings and real-world conservation efforts. These experiences support an emerging field of scholarship that posits video games as increasingly powerful affective systems that can bring people's thoughts and feelings closer together through the circulation of meanings and players' subjective connection to them. This chapter concludes by offering several provocations, drawn from the team's practical experiences, for those who want to make meaningful contributions toward sustainability through game development.

DOI: 10.1201/9781003512400-21

KEY TAKEAWAYS

- Designing for emotional impact rather than mechanical efficiency can create more meaningful and resonant gameplay.
- Aligning victory conditions with what players should care about—such as restoration over expansion—requires breaking away from entrenched design assumptions.
- Meaningful consultation can reveal unexpected design opportunities, shaping both mechanics and thematic clarity.
- Integrating real ecological principles into gameplay fosters learning, reflection, and a tangible connection between in-game actions and sustainability.

INTRODUCTION

Terra Nil is a hybrid strategy puzzle relaxation game where a player restores barren wastelands into a diverse set of thriving ecosystems. It borrows from the City Builder Games (CBGs) genre in many of its mechanics, where structures are created to simulate complex systems with many "interconnected and interdependent parts," introducing "geographical patterns and processes."[1] Breitenschaft argues that CBGs, as cultural products, have inspired "new generations of city planners, traffic engineers, and urban theorists" through their formative exposure to urban planning and development.[2] From a pedagogical standpoint, these games hold a powerful position in the shared imaginary[3] of what geographical planning entails. However, these games are also often marked by their ecological blind spots.

Unlike many CBGs, *Terra Nil* does not contain exploitation of resources, colonization of land, or unfettered expansion. It's a finite game—as developer Sam Alfred states: "Once you've restored an ecosystem, you move along."[4] But beyond simply tweaking mechanics, *Terra Nil* fundamentally rethinks what a game about land and development can be. By centering restoration, care, and departure rather than ownership and accumulation, it resists the dominant logic of control and extraction that has shaped not only CBGs but much of game design itself. This shift is not just mechanical but philosophical, requiring a break from long-standing assumptions about progress, success, and the role of the player in shaping virtual worlds.

In the following sections, the positionality and theoretical approach are outlined by delving into the development history of *Terra Nil*. This chapter demonstrates how player feedback became a central tenet of the design philosophy, shaping game mechanics, art, and sound throughout the development process. We also consider how the title relates to larger debates around real-world sustainability issues. Throughout and in the conclusion, a series of provocations are offered, derived from the development team's practical experiences, for anyone who wants to make meaningful contributions toward representing sustainability through game development.

APPROACH

This article was written by members of the development team behind *Terra Nil* and employees of South African indie games studio Free Lives. As such, it is important to counteract the "god trick"[5] and acknowledge that this chapter is hugely influenced by situated knowledge and embodiment and thus is naturally imbued with the developers' biases and subjectivities. The bulk of this chapter is an expansion of Sam Alfred's 2023 GDC talk "Developing '*Terra Nil*': A Strategy Game About Nature, Not Expansion"[6] and Jonathan Hau-Yoon's presentation "The Art of *Terra Nil*" delivered at Game Camp France,[7] along with additional consultations with the core development team.

In this chapter, "affect" is an important aspect in the development process of the game. In the introductory chapter of her book "playing with feelings," Anable cites a 1983 Electronic Arts advertising campaign that predicts that video games would become the "emotional touchstones of the future" a means to "bring people's thoughts and feelings closer together."[8] She makes the case that video games and affect theory are related historical and intellectual projects stemming from a long history of using computer models to "rethink the relationship between thoughts and feelings in human (and machine) consciousness."[9]

From the moment a tablet, phone, console, or computer is turned on to engage in a chosen game, the player becomes immersed in the mechanics, rhythms, colors, sounds, textures, and tones of that game. It requires playing by its rules and operating within its constraints. Games not only cognitively embed the player in their worlds but also *the world* at large—the relations and interfaces of social and political narrative, imaginary, ideology, aesthetics, and corporeal realities. Thus, importantly, Anable argues, following the lessons of Raymond Williams,[10] that video games "must be understood as more than just ideological training grounds for capitalism," but as "beacons for social change."[11]

In order to explore this concept, Anable puts forth a theorization of affect that, in its most basic understanding, posits affect as the "forces that inform our emotional states" influenced by the work of Eve Kosofsky Sedgwick, Lauren Berlant, Sara Ahmed, and Sianne Ngai.[12] In such an understanding, affect shapes objects and subjects as they meet, as means to reinforce, or counter, dominant, individual, normative, or progressive ideas. In this way, affect can articulate "underrepresented feelings" through cultural expression, objects, and ideas, to bind bodies through shared or similar experiences. Video games can thus become an interface for contemporary shared imaginaries of sustainability as "imagined forms of durable social and natural life that can stretch into the future."[13]

In the next section, details about the development trajectory of *Terra Nil* are explained, demonstrating how centering player affect became a tenet of the design philosophy.

TERRA NIL: CONCEPT AND DEVELOPMENT

It is important to note, right off the bat, that *Terra Nil* was not designed as a so-called serious game or a game that was intended to change the world. The original prototype of *Terra Nil* was created as a part of the 48-hour Ludum Dare jam in October 2019, for which the theme was "start with nothing." At the time, lead developer Sam Alfred was largely influenced by the game "Islanders,"[14] citing the fact that it was "a minimalist city builder, that is really relaxing and doesn't have a lot of time pressure."[15] He wanted to create something with these CBG mechanics but was instead inspired by nature. Thus, "instead of building a city or a factory, you were building an ecosystem."[16] The original version had six buildings, which constituted a mixture of real-life scientific technologies and "very hand wavey game magic"[17] to enable the player's goal of covering 50% of the landscape in greenery.

The core gameplay loop revolved around an empty "wasteland" map, featuring a smattering of rocks. The player could use these rocks to place wind turbines, which in turn could power "toxin scrubbers" which would clean the soil. This fertile soil would then be able to host greenhouses, which allowed greenery to grow. Having a sufficient area of the map reclaimed by greenery was the game's win condition.

However, there weren't enough rocks on the map to fully terraform the landscape, which led to the introduction of an "excavator" that would carve straight riverbeds. Those rivers could then be filled with water using water pumps, which would green the riverbanks, and the rivers would be sites for "calcifiers" which could turn adjacent greenery into rocks, which could then be used to plant further wind turbines. The game performed very well in the jam and placed 4th overall, as well as 1st for graphics.

While Alfred was surprised at the first placement for graphics—the art was "programmer art" and he didn't think it deserved a first place—he theorized that

> the idea is that players were just enjoying bringing life back to the barren wasteland. Players really like coloring-in books, players like turning something dirty into something clean. You just have to look at something like PowerWash Simulator to see this happen.[18]

The success of the jam version led Free Lives to support further exploration of the concept.

Applying Real-World Inspiration behind Mechanics

The team found that the core loop was mechanically sound and that attempts to make it longer or make subloops were not enhancing the experience. These strategies for expansion ignored what people really liked about the game, which was the satisfaction of bringing life back to the wasteland. This led the team to ask "what if we doubled down on the nature aspect?" The central tenet of the design philosophy behind *Terra Nil* would focus on placing nature first. The team took inspiration from their surroundings in Cape Town, South Africa, one of the most diverse ecosystems on the planet. As Alfred muses, "Where *Terra Nil* was conceived and developed ended up being quite important to its design."[19]

Looking at nature, Alfred argues that "nature is not homogenous"—while the original game had elements such as water, greenery, and soil, there wasn't any diversity in their composition. The team studied the complex particulars of wetlands, jungles, arid regions, and so forth, as their inspiration—and decided that their next steps would focus on biodiversity.

Observing the cyclical nature of Cape Town's fynbos, the team used a unique example from their own city—the Protea—as inspiration for a game mechanic. Alfred explains: "We have this great example of life and death on our doorstep."[20] Proteas are pyrophytic flowers, meaning they require fire to germinate. After the flower is pollinated, they will close their petals and wait (sometimes for years) to burn, so that they can release their seeds. When the grass below the Protea bush is burnt away, it gives the seeds a better chance to establish and grow. The ash, in turn, is a source of nutrition.

Guided by this real-life example, the team challenged players to use destructive means to introduce biodiversity into their goals. After the initial reclamation phase, when the player had reached enough greenery and water, the game prompted them to introduce biodiversity to the landscape: the map would start with a couple of dead trees, on which the player could place beehives, and those beehives would trigger the creation of fynbos from the grasslands around them. The player could then use a "solar amplifier" to create a controlled burn, which would create ash and building husks that, in turn, would grow forests from the nutritious ash. At the same time, greenhouses could be upgraded into "hydroponia," which could create wetlands, and "bio-domes," which could create arid plants on raised ground. All these things together became the goal of the second phase of the game—to improve biodiversity.

These additions had significant implications for the design—*Terra Nil* became a game about balance. As Alfred explained: "Nature is not about infinite growth, its ecosystems find a natural equilibrium. If you use tiles for a wetland, you cannot create fynbos there. The game is finite—if you run out of land because you haven't created balance, it invalidates the [design] philosophy, which is that nature would inspire the mechanics."[21]

In this way, the team designed victory conditions and mechanics of limited player agency and balance to emphasize the value of nature and biodiversity.

Prioritizing Affective Experiences

The team soon found themselves at a crossroad—while their game centered on restoring nature, the landscape was becoming cluttered with structures. As Alfred points out: "the thing about making a game about taking actions by building things, is that building things is your main *verb*."[22] To maintain this core mechanic while preventing the terrain from resembling a typical CBG cityscape, they introduced a recycling phase. In this phase, players constructed specialized buildings—"recycling beacons" or "recycling silos"—to dismantle previous structures. These materials were then transported via a loading port to build a rocket, allowing players to leave behind an untouched landscape. Alfred described this final phase as "not as mechanically deep as the previous two stages—it was a little more of a victory lap."[23]

This decision reflected a deliberate shift toward prioritizing affective experience over strict mechanical complexity. Rather than adhering to conventional CBG progression—where expansion and optimization are key—the team emphasized restoration as the ultimate goal. The recycling phase not only provided narrative closure but also reinforced the game's environmental message, allowing players to witness the full cycle of regeneration. By subverting traditional CBG expectations, Terra Nil offered a meditative, almost ritualistic conclusion—one that resonated emotionally, rather than mechanically, with players.

The team launched the free biodiversity prototype on itch.io in January 2021 and added the recycling feature in June of the same year. The additions really resonated with players and a host of YouTubers covered the game:

> Watching people play, watching YouTubers, reading comments, we realized that what really resonated with people was the overall experience. It's the fantasy of cleaning up the wasteland. The puzzle mechanics were good, they were *fine*...But without the *experience* they wouldn't have been sufficient. I don't think anyone would have been playing the game if it wasn't for the feelings they were getting.[24]

Hau-Yoon reflects on a particular comment that moved him, where a player mentioned the game had "broken through her antidepressants and had made her cry."[25]

> I think that in the context of imminent climate disaster, it's easy to get crushed by anxiety or apathy. I think it's a really good thing to create more media that exercises our imaginations and shows a hopeful future. Maybe it's only science fiction, but I think it's important to have some hopeful pieces of media in our collective imaginations. And that hopeful aesthetic is something important in our artistic choices and in our music. [26]

The studio decided that the game warranted further development beyond the prototype phase and put it into full development. Starting with a fresh codebase, the developers decided that they would use the aspect that generated the most affect with players—the experience of healing and restoring nature—and use that as a proverbial North star to build out the other pillars of the game's design.

Consulting Communities

Throughout the development process, the team engaged with a captured audience to guide their process in art, music, and mechanics.

While the game's earlier prototype consisted of a cheap and easy-to-replicate pixel art style, the team lamented its "jagged edges" finish.[27] As lead artist on the project, Jonathan Hau-Yoon, explains: "While the pixel art had appeal, pixel art can appear rather harsh and the screen can be rather hard to read, especially when one of our goals was to show life everywhere. The game also had a lot of text and UI, another area where pixel art can be a strain."[28]

The team's initial art direction was very inspired by Studio Ghibli's retrofuturism and steampunk aesthetics. This proposed influence offered a shorthand to a visual lexicon which audiences recognize as thematically tied to human-nature tension, environmental activism, and spiritualism in nature.[29] The team switched to a 3D environment that would allow a soft naturalism in the space.

Experimenting with a Ghibli approach, however, had some mismatches with the core themes of the game: "While an animated movie might want to push its characters into the foreground, we found that the buildings would become too prominent. In *Terra Nil*, even though you play the game by creating structures, it is always in service of your making changes to the environment. So, the environment is […] more like a main character, so it shouldn't be blended into the background."[30]

The team then created a diorama of a proposed style—with painted backgrounds and some more stylized fictional buildings—to run past the community. Unexpectedly, this prompted mixed responses—most significantly, the players lamented that they missed the iconic wind turbines. This was a valuable insight to the development team—that the players wanted a level of realism from the game. The wind turbine represented a real-world anchor that allowed people to understand and relate to what was happening in the game.

Hau-Yoon reflects on this moment and how it was a blogpost written by one of their players, Miguel de Dios,[31] which convinced them to pivot. Dios writes: "It's clear that the damages we have dealt the environment can only be undone with the same strength we used to build entire societies from scratch. To restore a plundered earth, we must employ industries, not individuals."[32]

The team decided that they would embrace the real-world inspiration behind many of their inventions, instead of emphasizing the otherworldly and fantastical, à la Ghibli. This, they decided, would better embody action against climate change.

Moving into a mixture of 3D modeling and digitally painted surfaces offered the art team softer curves—and afforded a level of realism that could be brought into the representation of this imaginary world. From a technical level, this meant a 3D approach with a locked-in isometric camera. Most of the buildings were modeled in 3D and textures were then projected to maintain a soft painterly effect without expending unnecessary time texturing an entire 3D object. Dynamic shaders that animated noises over time were used to add natural and subtle movement to the water and foliage. Artists Marcelle Marais and Kane Forster modeled and animated the animals, respectively.

For music, Hau-Yoon suggested that they use the free online catalogue of French artist, Meydän, during the earliest versions of the game. The soundscapes consisted of deeply emotive piano music. Their choice of soundtrack proved immensely popular with players, who would often request the tracks. The studio subsequently decided to commission Meydän to produce an original score for the game. The use of physical instruments added an analogue dimension to the soundtrack.

For sound effects, Free Lives' in-house composer Jason Sutherland also placed naturalism and technical realism at the core of his scoring. For the buildings, he says: "They sounded kind of metallic and old. Like a bit steampunk, heavy industrial. They had this very involved mechanical sound to it. I used steam and cranks and stuff."[33] As Anable argues "listening enhances looking,"[34] and through this sound design the team created an aural landscape that could provide additional information about the process-ing of the individual building's functions.

Aiming to increase the interactivity of sound, the team also worked on a parallel ambience prototype, where the cursor in *Terra Nil* acted as a microphone to amplify spatial audio on the map—based on the placed vegetation and buildings: "Making the player feel like they were in the space. Like they could interact with the mouse cursor."[35] This allowed players to gain a closer "entanglement" with the game—creating an expe-rience of being immersed *in* nature, rather than manipulating it.

As development proceeded, the team realized that toward the end of a level, players would frequently have holes in their landscape that wouldn't be covered in greenery. To remedy this, they introduced a climate mechanic: once the local weather system was restored, it would trigger rainfall. This rain would fill in the remaining gaps on the map. What started as a mechanical device, became a core emotional beat in *Terra Nil*'s gameplay, a moment which Meydän scored as such: "It was almost like the game was saying to players—' you've done enough…nature will take it from here'."[36] As Alfred argues, this proved that the game "could also support moments of beauty, the features didn't all have to be mechanical in nature." Again, like the recycling feature, the team opted to prioritize affective experiences over optimal design, forgiving imperfections or even some poor gameplay choices on the side of the player.

The next iteration of the game debuted on June 16, 2021, at Steam Next Fest, the game was received with a huge positive response and had over 100,000 players that week. Instead of describing the appeal of the mechanics, a large portion of players described their affective experiences.

Release and Continued Development

Netflix acquired rights to the mobile version of *Terra Nil* and the team continued work-ing on the full game, now with a stringent deadline in mind.

Taking their lead from the Köppen-Geiger climate classification system—where climate zones are divided into five broad categories, and 16 subcategories of biomes on earth—the team set to work on creating "one level for each broad category."[37] The full game would therefore consist of a Temperate, Tropical, Polar, Continental, and Arid region. Each region had brand-new mechanics, biomes, animal species, and climates. The Arid region was initially cut due to time constraints.

The team also added difficulty settings to allow players to control their preferred experience of the game. The team spent a great deal of time working on a balance between the relaxing elements and the challenging elements of the game and represent-ing those through difficulty choices.

Furthermore, the team added an "appreciate mode" which features pans and close ups of the map that the player had created—giving them an opportunity to reflect on their achievement. Alfred said: "We wouldn't have bothered with a feature like this, if we didn't listen to the game."[38]

Terra Nil was released on March 28, 2023, for Windows, Android, and iOS, with the switch version launching in December of the same year. The game sold around 250,000 units in the first year on Steam, a secondary platform to the huge reach of

Netflix on mobile, and Switch. The game has been nominated for a multitude of categories in some of the world's biggest game awards, and critics have been largely favorable.

After taking a brief hiatus, the team returned to work on *Terra Nil* later in 2023 with an external update team and released an update, titled Vita Nova, with expanded levels. The team decided to continue development on the Arid region in-house, which, as this chapter is being written, is their focus. The update is planned for release in 2025.

CONTRIBUTIONS TO SUSTAINABILITY

There is an emerging field of scholarship which posits video games as increasingly powerful affective systems[39] that can bring people's thoughts and feelings closer together through the circulation of meanings and subjective processes of identification. As Hau-Yoon argues:

> Post-apocalyptic depictions of the world dominate science fiction and many of these views are very cynical. And although I can understand this point of view, the stories we hear about corporate greed and such are certainly depressing—but when such a large part of our imaginations is taken up by these depictions of the future, it can make this future seem inevitable. *Terra Nil* is technically post-apocalyptic: The world starts as a lifeless wasteland, but the game gives players the agency to change things, agency we don't often feel dealing with global problems. The game doesn't prioritize amassing as much money as possible, there are no high scores, the map is not endlessly large, and you cannot expand indefinitely. Your goals are using your limited land and limited resources to balance ecosystems and animal habitats and create life. Against a backdrop where we're often told that we're past the point of no return, that recycling is ultimately pointless, that anything we can do individually is a tiny drop in the ocean against the hundred or so companies that cause 70% of global emissions. That kind of attitude is intended to inspire urgency, but for me it caused hopelessness.[40]

In this way, *Terra Nil* inspires a new hopeful imaginary of the future. By foregrounding real-life inspired industrial-level solutions, the game argues for the potential of technologies, strategies like rewilding and conservation efforts, to combat the fallout of climate change. It offers an everyday entanglement with a wide array of subjects that both teach about, and explore, examples of contemporary environmentalism—through an "affective assemblage" of "intimacy and entanglement."[41] In this way, *Terra Nil* offers a beacon for social change by countering the dominant ideologies surrounding our shared future. It offers an interface that can galvanize communities in their want for a more sustainable life on this planet.

The team realized the power of such a galvanized community and offered players a concrete means to proverbially walk the talk. They approached the Endangered Wildlife Trust (EWT), an organization with its headquarters in South Africa, and offered to donate a percentage of the game's profits to their conservation efforts. The EWT, which was established in 1973, is dedicated to "conserving threatened species and ecosystems in Southern and East Africa to the benefit of all."[42] Up to date, *Terra Nil*'s donation to the organization has been more than one hundred thousand euros, making *Terra Nil* among the biggest donors in the organization's history.

CONCLUSION AND PROVOCATIONS

The development of *Terra Nil* provides a compelling case study for game designers seeking to engage with sustainability themes in ways that go beyond mere representation. Through iterative design processes, community feedback, and grounding their mechanics in real-world ecosystems, the team was able to craft a deeply affective experience centered on environmental restoration. The success of *Terra Nil* ultimately highlights that sustainability in game design is not just a question of thematic alignment, but also of how effectively the design choices support the intended emotional and cognitive experiences of players.

To further explore the implications of this approach, the following four provocations are offered:

1. **Prioritize affective experience over optimal design:** Traditional design wisdom emphasizes balance, efficiency, and the optimization of systems. However, *Terra Nil* illustrates that sub-optimal design can be in service of a more powerful experience. If every element in the game supports a specific emotional journey, designers can justify departures from standard practices in favor of greater resonance. The game's recycling phase, for example, was designed as a "victory lap" rather than a deep mechanical challenge, but this celebratory moment was important to complete the arc of restoration.

2. **Design victory conditions around what you want players to value:** If a game asks players to care about something, it should be central to how they succeed. In *Terra Nil*, greenery and biodiversity weren't just aesthetic elements, but the win conditions themselves, ensuring that players inherently valued restoration over expansion. However, achieving this shift required more than just aligning mechanics with intended values—it meant stepping away from entrenched design assumptions that prioritize ownership, optimization, and extraction. By disregarding conventional city-building paradigms and reimagining success as leaving no trace, *Terra Nil* invites players into a radically different experience—one that resists the logic of endless growth and instead fosters a sense of care, stewardship, and collective restoration. Designing for hope in this way demands more than tweaking mechanics; it requires a fundamental departure from the systems of thought that shape dominant game design traditions.

3. **Consult communities to illuminate unknowns:** The team's engagement with their community highlighted how meaningful consultation can reveal unexpected insights. Feedback that the game's buildings should look like "heavy machinery" provided an important turning point, allowing the developers to communicate the magnitude of climate restoration efforts more effectively. This underscores that consultation is not just a source of feedback but can also be a pathway to discovering hidden design opportunities and thematic refinements.

4. **Anchor mechanics in real-world inspirations:** Grounding game mechanics in real-world systems, as seen with the pyrophytic Protea flowers, creates a space for players to learn about and reflect on ecological principles. When done well, these mechanics allow players to explore contemporary conservation efforts and consider new possibilities for a sustainable future. By embedding tangible real-world inspirations within game systems, designers can create a bridge between fictional play and real-world action, fostering both understanding and hope.

Through these provocations, we challenge designers to reimagine how sustainability can shape their design practices—not as a constraint, but as an opportunity to create more meaningful, resonant, and impactful experiences.

NOTES

1 Breitenschaft, Bradley. 2016. "Gods of the City? Reflecting on City Building Games as an Early Introduction to Urban Systems." *Journal of Geography* 115 (2): 51–60.
2 Breitenschaft, Bradley. 2016.
3 Miller, Thaddeus R. 2020. "Imaginaries of Sustainability: The Techno-Politics of Smart Cities." *Science as Culture* 29 (3): 365–387.
4 Game Developers Conference. 2024. *Developing 'Terra Nil': A Strategy Game About Nature, Not Expansion*. April 16. Accessed September 19, 2024. https://www.youtube.com/watch?v=s30pjYV8aBM.
5 Haraway, Donna. 1988. "Situated Knowledges: The Science Question in Feminism and the Privilege of Partial Perspective." *Feminist Studies* 3 (1988): 575–599.
6 Game Developers Conference. 2024.
7 Game Camp France. 2024. *The Art of Terra Nil | Jonathan Hau-Yoon | Game Camp France 2024*. Lille, June 24–25.
8 Anable, Aubrey. 2018. *Playing with Feelings*. Minneapolis: University of Minnesota Press, 1–3.
9 Anable 2018, 5.
10 Williams, Raymond. 1977. "Structure of Feeling." *Marxism and Literature* (Oxford University Press) 128–135.
11 Anable 2018, 7.
12 Anable 2018, 10–11.
13 Miller 2020, 3.
14 Grizzly Games. 2019. *Islanders*. Berlin.
15 Game Developers Conference 2024.
16 Game Developers Conference 2024.
17 Game Developers Conference 2024.
18 Game Developers Conference 2024.
19 Game Developers Conference 2024.
20 Game Developers Conference 2024.
21 Game Developers Conference 2024.
22 Game Developers Conference 2024.
23 Game Developers Conference 2024.

24 Game Developers Conference 2024.
25 Game Camp France. 2024. *The Art of Terra Nil | Jonathan Hau-Yoon | Game Camp France 2024*. Lille, June 24–25.
26 Game Camp France 2024.
27 Game Developers Conference 2024.
28 Game Camp France 2024.
29 Gossin, Pamela. 2021. "Animated Nature: Aesthetics, Ethics, and Empathy in Miyazaki Hayao's Ecophilosophy." *Mechademia: Second Arc* (University of Minnesota Press) 10: 209–234.
30 Game Camp France 2024.
31 De Dios, Miguel. 2020. "Industries, Not Individuals Will Save the Earth." *Cutscenes*. November 15. Accessed October 6, 2024. https://cutscenes.substack.com/p/industries-will-save-the-earth.
32 De Dios 2020.
33 Sutherland, Jason, interview by Anja Venter. 2024. (October 6).
34 Anable 2018.
35 Sutherland, Jason 2024.
36 Game Developers Conference 2024.
37 Game Developers Conference 2024.
38 Game Developers Conference 2024.
39 Anable, 2018.
40 Game Camp France 2024.
41 Anable 2018, 76.
42 Endangered Wildlife Trust. 2024. "Endangered Wildlife Trust." *Endangered Wildlife Trust*. https://ewt.org.za/.

Empowering Communities to Define Climate Goals

19

Clayton Whittle

ABSTRACT

This chapter challenges conventional approaches to climate-focused game design by shifting from persuasion to empowerment. Drawing from original research and fieldwork, Clayton Whittle explores how player communities, cultural context, and social ecosystems profoundly shape the efficacy of environmental games. Through case studies and analysis, this chapter reveals how interventions that fail to understand players' lived experiences, motivations, and support needs often miss their intended impact. By advocating for community-informed goal setting, intentional support structures, and participatory design, this chapter reframes games not as tools to transform players but as platforms that empower them to act. The path to meaningful climate impact, it argues, lies in designing with—not for—communities.

KEY TAKEAWAYS

- **Empower, don't persuade**: Games should support motivated players rather than trying to change them.
- **Community matters**: Player behavior is shaped by the game's social environment.
- **Context is critical**: Designers must understand where players are starting from before setting goals.

DOI: 10.1201/9781003512400-22

- **Support is essential:** Poorly supported interventions can harm even well-intentioned players.
- **Think beyond gameplay:** Impactful design includes community, tools, and long-term support.

FROM A TO B

What does it mean to change the world? For what impact might any game developer strive? How can we measure that impact? How can we choose that impact's nature, scope, scale, geographic location, or any of a 1,000 critical choices that must be made to have an effective climate impact?

When training developers to create climate games, I make it salient that these questions are pivotal in understanding how to achieve their goals. I insist that they follow the mantra "aim small, miss small." I challenge designers to choose a specific variable to target and design their intervention with that goal in mind.

In these sessions, heavily guided by the context of designing a game to affect a player's behavior shift, I coach designers toward considering their audience. I use a simple method. I draw a "B" on the board. I tell the attendees that the "B" represents their behavioral goal. I then instruct the attendees to draw a line from "A" to "B."

The question is inevitable: "Where is A?"

The point then, blunt as it is, is that we cannot hope to enact change, to meaningfully transform any individual without understanding the starting point of said individual.

Within the design-oriented workshop, this is the point at which we continue to a user research element, a method of understanding that starting point. There is, however, an unspoken assumption that "B" is a desirable, impactful, or reasonable end point for the targeted player.

We cannot, however, be certain that this is the truth. We cannot know what it is that the individual player needs, what knowledge they may require to take more meaningful action, what issues impact their daily lives, or what attitudes they identify with. Even if we understand the starting point, the "A," to presume that "B" is the most impactful end point disempowers the player.

The folly implicit in the assumption of a "B" goal is best illustrated through a research paper I often reference, *We Be Burnin'*.[1] In this project, a group of researchers set out to build a summer climate education program. In doing so, they created an entire program, from start to finish, for a group of youths. In the course of the program, however, the youth began to indicate that they lacked interest in both the content and method of instruction. Their interest had been sparked, yes, but not in the way the research team had predicted.

Wisely, the team adjusted. They allowed the youth to pursue their own local investigation, to drive the content of the program by investing in issues that felt important to them, and to report on those issues through a medium that resonated. The research team made their resources available to the participants to reallocate and pursue their own goals.

The result was something the team deemed "community science experts," youth with scientific knowledge, emotional and social investment, and a systematic understanding of their local challenges and assets. Empowered thusly, these youth were able to make impactful and reasonable individual and institutional recommendations on improving resiliency in their area.

The *We Be Burnin'* project illustrates an approach to environmental and climate issues that, while increasingly common among environmental impact organizations, remains less recognized in the world of media-for-impact.

Back to B-sics

The approach with which we began this conversation, that of attempting to guide an end user toward the stated goal of "reaching B" can be considered a *climate resilience education intervention*. That language, though, can be loaded with an assumption that an intervention must be done "at" a user. In truth, though, an intervention can also be designed in collaboration with its end users/co-creators, taking into account their needs and the perceived cultural and social context.

The *We Be Burnin* example shows how a culture defined not only the "A" starting point, but the "B" ending point as well. The research team needed to reexamine their end goal and redesign it alongside the learners.

Games as Culture

The importance of the cultural context is equally critical, when we discuss games for impact, because players exist within both their traditional cultures (family, region, religion, etc.) and the culture of players surrounding the game itself.

What we learn from games is driven by the cultures around them.[2] We often call these cultures the "metagame." These spaces are inevitably tied to our understanding of how gamers interact with games. They can also be spaces of learning.[3] Initially, the perception was that those who were truly dedicated to any specific game may find themselves socially learning from the interaction within that community.

As designers, educators, and researchers, we have used the past decade to improve our practice in creating persuasive games. In truth, many steps have been taken in that direction, and we find ourselves increasingly certain of effective design patterns. But games, no matter how well designed, are moments in a lifetime of moments. They are a single drop in a stream of information which leads to a decision.

More to the point, they are inanimate and unjudging objects, in direct contrast to the community of individuals around the game, each of which is capable of exerting social influence through judgment.

Individual Action

I adhere strongly to the notion that the ultimate evidence of successful environmental education is an individual pursuing, of their own volition, pro-environmental action.[4]

But any environmental intervention designed to influence an individual, must account for not only how the individual receives the message, but how the community receives it! Now that we have come so far in our understanding of how to design incredible climate games, we must next ask ourselves: how well we are supporting our players in acting upon the messages in those games and in their own social communities?

In this chapter, I offer a perspective on how the socio-normative contexts through which players view green game designs impact their perceptions, exploring how they experience green games through the lens of their own cultures. I call on recent research completed in collaboration with the Playing 4 the Planet Green Game Jam to inform this work. This chapter strives to highlight a path along which the movement of climate games may yet make great strides and improvements.

Through the stories of two players, I hope to guide our conversation toward supporting the players of our games and seeing them in the same way that Learning Scientists have seen them since the introduction of sociocultural learning theory,[5] as active participants whose experiences must guide our designs, rather than our designs their experiences.

Functionally, in this chapter, I argue for an approach to green game design that is more holistic in nature, merging the concepts of in-game design, social media support, and intentional community management. I lay out an argument for the need for intentional community management and support as part of that holistic approach, drawing heavily on the toxic or non-supportive community experiences of players. I will accompany this with a process for community development based in climate action psychology, established professional community management practice, and community action research.

Through this exploration, I try to help game developers consider the crucial importance of the social environment of their players so that they can better define where their "B" is.

A History of this Work

The most appropriate place to begin is, perhaps, at the beginning, the goal of this research, this chapter, and this movement. I will state it plainly, as it appears to me:

I wish to discover how we might use games to empower individuals to enact change that creates a more sustainable world.

Lofty, I know, and perhaps preachy. However, without this lofty goal, we cannot contextualize the remainder of this chapter, cannot show what it is I believe the game industry is truly capable of or how they have not yet lived up to that potential.

More salient, though, is that we must ground ourselves in that goal. Too often, the temptation to fall in love with making great green games distracts from the goal of building a more sustainable world, and we really need to do both.

The Work

In a recent research project, I spoke at length with long-term players of a popular game. These players were established parts of the game's culture, members of a community. This game was in the middle of a campaign to encourage players to modify their eating habits to include less meat to reduce their carbon impact. Their behavior changes campaign included in-game elements and a social media campaign focused on vegetable-based diets.

I followed the experiences of two of these players in great detail, spending several hours speaking with them one-on-one and facilitating an in-depth journaling process, alongside my partners Trevin York and Amy Rodger.

These journals and interviews reveal much about the players' lived experiences and the context of their decision-making processes that efficacy research or click metrics simply cannot. Their experiences illustrate how the studio's focus on behavior shifts, rather than empowerment, failed to capitalize on the existing pro-environmental momentum of each of these players.

As a note, descriptions of game elements will remain intentionally vague, both to protect the studio's anonymity and to encourage the reader to see the story as applicable to their own work.

Shalesh and the Need for Support

Shalesh, a woman in her early 1920s living with her family in Australia while attending university, was already familiar with vegetarianism. While her family still ate meat on occasion, they did so in a way that respected her father's wishes and beliefs as a vegetarian.

Shalesh is a perfect example of a player who needs to have her mission and identity supported by her favorite game. Over the course of six weeks, Shalesh made 42 journal entries, each detailing her thoughts around the game, the game community, and her own dietary choices or events of the day.

Though our research sought to understand the player's reaction to the dietary change campaign, it became abundantly clear that Shalesh was more interested in the community reaction than the intervention itself. In Figure 19.1 (note that details which may identify the game or participant have been removed), a typical entry, Shalesh describes her play for the day. Her focus, however, documents the engagement with the campaign on social media, thoughtfully discussing how the community may or may not understand the sustainable mission.

Shalesh continues her focus on the community of players throughout. She begins to include a rudimentary analysis of social media and community responses. In a separate journal entry, Shalesh's notes specifically call out that the community response is primarily negative and even "against vegan food." Specifically, she writes that she was responded to with "mainly negative comments against vegan food."

Her journals become even more analytical, calling out the percentage of responses that are either negative or positive. She notes frequently that the negative responses outweigh the positive ones. She laments how, in her perception, the people who react to the social media campaign posts by the game studio have "missed the point that a vegetarian diet is always an option," and treat the campaign as "vegan propaganda."

Indeed, our own analysis of the social media campaign supports this conclusion. Even when posts were indirect and asked for followers to "submit vegetable-based recipes," responses leaned toward intentionally antagonistic. I will not quote social media

Thursday 5th September

played on bus to and from uni

Engaged with ▮▮▮▮▮ on social media

○ — liked + commented on posts

○ ▮▮▮▮▮ x meme — ♡ 7k

▮▮▮▮▮ post — ♡ 1.6 k

— Daily High Score

— Back to School Challenge

The ▮▮▮▮ osts real money which younger generations have less access to, and therefor don't prefer those options. There was also some conflict and confusion as to why a video game would be promoting environmental impact given video games themselves don't really have a good impact on the environment themselves. I get that it might be hard for some people to understand:
It's easier, makes more sense, and is actually possible to promote positive environmental impact through video games rather than making an impact by stopping people from playing video games altogether.

FIGURE 19.1   Journal entry by Shalesh. Identifying markers removed.

posts here, but many took the opportunity to express dislike for vegetarians as people, suggest bacon as a vegetable, or simply insult the social media manager.

In interviews, Shalesh laments the experience and, despite a generally positive disposition, shows few signs of finding any value in the campaign. Naively, I take this as a sign that Shalesh simply does not need more affirmation or guidance in vegetable-based dietary planning. This view is shattered when Shalesh excitedly journals about her joy in seeing a vegetarian character in the game. Her language expresses not only joy in representation but also being "glad to see that (the character) was a well-liked character."

A short sentence, a notation. But, at the same time, a critical and explicit desire to have her own identity (or at least one she empathizes with) represented and accepted, even in the form of a digital character.

Should we be surprised?

Shalesh was looking for acceptance, for her gamer community to show her the support that her family did. Indeed, she actively searched for "supportive comments" and documented their occurrences in social media replies. Instead, she found an overwhelming number of replies that made her feel small, unseen, or unwanted.

Kalak and the Need for Tools

Kalak, a recent university graduate in Vietnam who had just moved back to his hometown, thought constantly of the environmental impacts of his behaviors. He argued with friends about their choices and kept a small garden as he once heard that plants cleaned the air of pollution.

Kalak represents an entirely different possibility: an individual so hungry to learn about environmentalism from games that he could individually skew a study on the topic with his responses.

Kalak says, "saving the environment is like my personality." Kalak is also deeply aware of and passionate about his ability to learn life skills or traits from video games. He touts how a character arcs other games encouraged him to pursue inner peace through a meditation practice and inspired him to move back to the country with his family. Kalak even keeps a little tree garden on his roof, having learned of the value of trees for the carbon cycle from a video game.

In short, Kalak is a perfect player for a game intending to teach about sustainability or environmental issues. Indeed, when Kalak heard about the event introducing vegetable-based diets, he was beyond jovial, literally bouncing in his chair during the initial interview.

Over the course of the six-week campaign, Kalak chose to submit his journals via video or audio recording. An avid streamer, this format felt comfortable to him. His initial excitement soon passed, though, as his entries moved toward painful experiences.

By the midway point of the project, it had become clear that Kalak was battling a depressive experience that was deeply connected to his engagement with the vegetable-based diet campaign. Kalak had, he reported, thrown himself full sail into the concept of reduced meat. He had done so in a way that was causing issues within his social life, though. Kalak's friends were not all supportive of the decision, and his journal entries mentioned tension or arguments among friend groups as his diet caused him to either feel ostracized or to frustrate others.

In later entries and interviews, Kalak reports significantly reduced bonds with his friends. There is even tension within his family, despite their efforts to support his mission. In our final interview, we ended early, as Kalak was simply overwhelmed by emotion and I could not, in good conscience, finish the project.

I would not claim that Kalak's experiences, stresses, and emotions were directly and uniquely caused by his sustainability pursuits. However, Kalak himself made clear that the social issues around sustainability strongly contributed to rather than alleviating his overall emotional stress.

Why Did We Fail

The vegetable-based diet behavior change campaign failed these players. Starkly and inarguably, it instigated behaviors that drove two greatly motivated individuals away from the intended choices.

Kalak was left feeling socially isolated. Isolation can lead directly to feelings of depression.[6] Shalash was made to feel ostracized and othered by the community, utterly undermining her pursuit of a sense of communal efficacy and acceptance/normalization of her way of life, both key components in long-term commitment to sustainable behaviors.[7] [8]

I do not argue that the campaign by this game did more harm than good. I doubt that is the case. However, I will argue that this campaign did far less good than it could have.

Kalak, more than anything, needed tools to empower his desire to teach and influence his friends about environmental issues. Instead, he was given yet another talking point to crusade about, yet another spike to drive between himself and his friends. Shalesh needed acceptance but was instead made the indirect target of vitriol and ridicule.

Acceptance.

The ability to convince and inspire.

These are not things that come easily. They cannot be taught by a single game and certainly not through a short in-game campaign.

More to the point, these needs represent a longitudinal problem. If games wish to enact pro-environmental behavioral shifts with real impact, then teams designing pro-environmental campaigns or messages must consider how the player will take that message or learning into the world and how action will be supported in the long run. We cannot realistically expect individuals to act without a support system. This means that it is a necessity to consider the social context of our players if we want them to be able to act!

How then can studios that wish to design similar interventions achieve their goals?

First, let us ask how these studios traditionally approach individual change.

Change Approaches through Design

Frequently, when impact-oriented games are discussed, we speak of the change or transformation they create. Indeed, one of the most influential texts on the topic is called *The Transformational Framework*.[9] Too frequently, though, the target of transformation is not the world, but the player.

There are four primary variables we rely upon to predict pro-environmental behavior: knowledge, attitude, perceived self-efficacy, and hope. Each, in its own way, presents itself as an outcome of a philosophy of climate activations that conceptualizes "change" as changing the player.

Many activations wish to "change the attitude," or "change the knowledge" an individual has. They represent an assumption that the individual must be changed to fuel climate action. They imply that, in the context of games, the player is not already prepared and motivated to act. Most critically, they strive to change the player, not the world.

These behavioral interventions are not without merit, and, in fact, I have argued strongly for refining our design methods for these types of games in the past.[10] While their specific theoretical basis or foundational theories of design vary widely, games that seek to change behavior of a player usually rely on a theory of change founded on addressing one of the four primary predictive variables of pro-environmental behavior: knowledge, attitude, hope, or perceived self-efficacy.[11]

Individual researchers or designers may use different vocabulary or predictive models, but these four variables are generally at the heart of behavior interventions, in games or other media, or educational interventions. Because of this, they warrant, at least, a brief explanation.

Cognitive psychologists, specifically those in the realm of climate psychology, generally agree that the four concepts mentioned above are reliable predictors of pro-environmental behavior. Should an individual be possessed of an *attitude* that drives them toward wanting to take action,[12] *hope* for a positive climate future,[13] *knowledge* necessary to take effective climate action,[14] and *perceived self-efficacy* that their individual actions can make an impact,[15] they will take pro-environmental action. This is an oversimplification, but one that illustrates the requisite components.

Interventions that aim to change environmental behavior frequently target one of these four variables, assuming that an appropriate change of state in one variable will increase likelihood of pro-environmental behavior.

My work on the *Environmental Game Design Playbook*[16] describes these theories and provides detailed examples.

The Process of Empowerment

Up to this point, this chapter has focused significant energy on illustrating a problem. This problem, as I have laid out, is one of incomplete in-game solutions and a philosophy of using games to change players to fit a designer-defined conceptualization of effective climate action.

This approach risks producing games that fail to achieve their climate goals. It is convenient for a studio design team to identify a climate goal, provide a solution, and use in-game or near-game interactions to briefly educate the player on the climate message. However, this approach risks a quick drop off in commitment to action, as it does not account for the player's own context and does not create support structures for pursuing longitudinal action.

So long as the development team sees themselves as an entity that persuades, educates, or inspires action, they must transform the player from their current state to one more appropriate to the final climate action goal. This perspective is incomplete.

There is a place for games to educate or persuade, but there is also a place for games to empower those who already possess the motivation to act. Herein, we can see games not as the spark to action, but as the tool employed by the gamer.

Through this perspective, we strive not to transform the player, but to empower them. In doing so, we align with more recent and robust practices across climate education and the learning sciences to give agency to those who are most capable of imagining and enacting meaningful climate impacts.

Communities for Action

Engagement with an environmentalist-minded community can significantly increase normalization of pro-environmental attitudes along a litany of pro-environmental actions and behaviors.[17] Communities can also provide opportunity for increased perception of self-efficacy as new members can be guided through personalized and skill/resource-appropriate challenges to increase their confidence.[18] In addition, communities consisting of knowledgeable individuals can help support the development of localized and actionable knowledge by responding to individual needs.[19] [20]

More to the point, communities allow for adaptive prioritization of goals and solutions, letting the community members identify the problems that are significant to them, suggest solutions, and select the most relevant solution for the context. Communities also enforce social boundaries and acceptable behaviors in way that anonymous or pseudo-anonymous social media platforms do not.[21]

Communities capable of adapting to the needs of players offer a promising path toward actual empowerment of individual players. Revisiting the stories of Kalak and Shalesh, it is not hard to see how their stories might have turned out differently, had the game studio made efforts to provide communal support rather than generalized solutions. An intentionally moderated community might have offered a place where Shalesh could explore ways to merge her and reconcile her dual identities as a gamer and a vegetable-heavy eater rather than having these two identities represented as mutually exclusive. A guided community might have provided Kalak with a place to express his concerns and struggles regarding his relationships and desires to persuade his friends and family.[22]

Most importantly, agnostic to the stories of Kalak and Shalesh, a community would have provided a space for players to raise their own concerns, voice their struggles, and interact to find solutions. Kalak and Shalesh are two examples, but the many struggles of other players also deserve to be heard, voices that were silenced by ignorance rather than malice.

Start with Engagement, Not Development

As with the original workshop story, the development of community-empowered interventions begins with the defining of a "point A." Before determining how a pro-environmental design or campaign should look, determine the current state of your game's community, including their desires, needs, and resources.

Engage with moderators, influencers, and Key community Members. Initiating engagement with pivotal community figures is essential for understanding the community's dynamics and fostering an environment that is supportive of the message and mission the team is pursuing.

Start by identifying key figures who represent diverse perspectives. It is important to pursue perspectives that are representative of the community, not just the loudest voices or the ones most likely to agree with a pro-environmental campaign.

With these individuals, seek to understand the community's current engagement with environmental themes. This process begins with a pursuit of critical information.

To What Extent Will Pro-Environmental Messages Encounter Resistance

Not all members of a player community will be open to pro-environmental messaging. As we have seen in previous research, this may be due to explicit political or scientific beliefs. It may also simply be caused by a desire to keep a community status quo and prevent a perceived change in messages and topics within the community.

If strong resistance to initial pro-environmental messaging is possible, consider the following steps.

1. **Start small:** Implement incremental changes and do not enact dramatic shifts in messaging or communications between the staff and the community.
2. **Promote inclusivity**: Ensure that new discussions, spaces, or activities welcome a wide range of voices and perspectives. Encourage players to engage with sustainability without alienating those uninterested in the topic.
3. **Empower allies**: Recruit supportive members to champion the changes and help normalize new initiatives.
4. **Framing**: Frame sustainability as an extension of the game's core themes, such as teamwork, problem-solving, and creativity, to foster organic interest.

What Climate-adjacent Passions Exist in the Player Community

Identifying existing climate-adjacent passions, trends, or interests within the player community can greatly simplify attempts to create either in-game or community-based pro-environmental campaigns. Identifying these passions can help teams quickly locate overlaps between the game's themes, sustainability issues, and the interests of the community at large. These overlaps are not always as obvious, and performing this initial analysis can increase the likelihood that pro-environmental campaigns will resonate with the community. Fayolle (Chapter 8) has this as a part of his model for considering your approach in your game as well.

What Current Activities or Approaches to Climate Activism Being Pursued by the Community

If your community has a strong pro-environmental element, there may be solutions already being pursued by community members. Understanding what these are can help your team to identify the "point B" that is so critical to developing community-led pro-environmental action.

Seek to understand what approaches, if any, the community pursues to combat the climate crisis. Are these solutions individual or communal? Are they perceived to be effective? What resources are available to the community to pursue these actions?

By identifying these activities and processes, we accomplish two critical goals. First, we identify community priorities. Second, we identify available resources and gaps in those resources. These resources might include educational resources, external organizations, third-party apps, or anything else that community members use to pursue their climate-adjacent goals.

In fostering a thriving and sustainable gaming community, it is essential to recognize and build upon the strengths that already exist. Players often generate innovative ideas, projects, and solutions that reflect their creativity and passion for the game. By magnifying these contributions, providing critical resources, and forming partnerships with established networks, you can create a collaborative ecosystem where players feel empowered to take ownership of their impact.

Once the team knows the priorities and resources being utilized, a strategy to support or supplement these resources can be developed.

Prioritize Creation over Consumption

Once analysis and initial engagement have been completed, there can be a temptation to take back control over a community-based campaign. However, the more the community is encouraged to actively create and participate, the more likely the community will internalize the campaign. Empowering players as creators builds collective knowledge and deepens their engagement both with the community and their own environment.

Projects like in-game challenges, modding competitions, or collaborative storytelling can encourage active participation. The vocalizing of challenges that may seem individual (such as Kalak's social struggles) can also spark innovation and support for new strategies for dealing with challenges.

Organize and Elevate

Even in organized communities, members may lack the resources or experience to organize their own activities. Drawing on learnings from your initial community engagement, work with influential community members to plan and organize community-based solutions. Provide access to resources that may be lacking in the community and assist community leaders in creating structure around solutions.

For example, this may include helping to establish clear guidelines for community content, facilitating a community-led hackathon, or offering rewards and showcasing opportunities for community member successes.

Keep in mind that leadership in community-based approaches should focus on guiding rather than directing. Empower players to take ownership of the space to foster a sense of collective responsibility and prioritize the needs and challenges of the players rather than the developers. The ultimate goal should be the shifting of organizational and leadership responsibilities to the community itself.

Provide Resources to Empower Players

Offering access to knowledge bases, experts, and tools empowers players to develop their ideas and projects, enhancing the community's creativity and innovation. Providing access to knowledge bases, experts, and tools empowers players to develop their ideas and contribute meaningfully to the community, fostering creativity and innovation.

However, identifying the most beneficial resources and ensuring their accessibility to all members, regardless of background or experience level, can be challenging. To address this, forming partnerships with experts, organizations, and mentors can provide valuable guidance through workshops and structured support. Additionally, creating a well-organized, regularly updated repository of resources ensures that community members have easy access to the information and tools they need, allowing the space to evolve based on participant feedback and emerging needs.

Partnering with organizations, influencers, or other game communities focused on sustainability can expand your community's reach and impact, introducing new perspectives and resources and leaning on existing expertise to provide established solutions instead of requiring your community to create their own from scratch.

Building a Better Community

From a functional standpoint, there are certain process steps that we can take to encourage the development of a supportive community. I will not argue that these steps are universal, only generally applicable to the majority of situations. And, as is often the case, the reader must filter this process through the lens of what is possible in their development context, taking into account the arguments I have made for the immense value of community management.

This approach is grounded in practical experience, collective best practices, and verified climate psychology research. It is informed significantly by the work of Gee and Hayes,[23] as well as existing best practices of community management in games.

The process should not be taken as a replacement for a complete process to community management. Nor is it meant to imply that teams interested in green games or pro-environmental campaigns in games have not already explored these steps. Much to the contrary, these insights are informed by observations of successful community management practices.

Rather, the process below is intended to help refocus and merge our community management practices and green game design practices to ensure that the focus remains on impact.

At the start of this chapter, I relayed a story. Regardless of the game being designed or the climate issue being focused on, the lesson of that story remains true. The mission of a green game campaign is to have a climate impact. To have a true impact, a team must be humble in seeking to understand their players, ensuring a true understanding of "point A" and "point B." It is only through this understanding that a team can seek to support and empower lasting transformation.

NOTES

1 Angela Calabrese Barton and Edna Tan, "We Be Burnin'! Agency, Identity, and Science Learning," *Journal of the Learning Sciences* 19, no. 2 (2010): 187–229, https://doi.org/10.1080/10508400903530044.

2 Yong Ming Kow, Timothy Young, and Katie Salen Tekinbas, "Crafting the Metagame: Connected Learning in the Starcraft II Community," *Connected Learning Working Papers* (2014): 1–46.

3 Kurt Squire, "Designed Cultures," in *Games, Learning, and Society: Learning and Meaning in the Digital Age*, ed. Constance Steinkuehler, Kurt Squire, and Sasha A Barab, 1st ed. (New York: c, 2012), 11–31.

4 Hollweg, Karen S., Jason R. Taylor, Rodger W. Bybee, Thomas J. Marcinkowski, William C. McBeth, and Pablo Zoido. "Developing a Framework for Assessing Environmental Literacy," *Washington, DC: North American Association for Environmental Education* 122 (2011).

5 V. John-Steiner and H. Mahn, "Sociocultural Approaches to Learning and Development: A Vygotskian Framework," *Educational Psychologist* 31, no. 3–4 (1996): 191–206.

6 Shuai Zhu et al., "Association between Social Isolation and Depression: Evidence from Longitudinal and Mendelian Randomization Analyses," *Journal of Affective Disorders* 350 (April 2024): 182–87, https://doi.org/10.1016/j.jad.2024.01.106.

7 Philipp Jugert et al., "Collective Efficacy Increases Pro-Environmental Intentions through Increasing Self-Efficacy," *Journal of Environmental Psychology* 48 (2016): 12–23, https://doi.org/10.1016/j.jenvp.2016.08.003.

8 Dian R. Sawitri, H. Hadiyanto, and Sudharto P. Hadi, "Pro-Environmental Behavior from a SocialCognitive Theory Perspective," *Procedia Environmental Sciences* 23 (2015): 27–33, https://doi.org/10.1016/j.proenv.2015.01.005.

9 Sabrina Culyba, The Transformational Framework: A Process Tool for the Development of Transformational Games., 1st ed. (Carnegie Melon University, 2018), https://doi.org/10.1075/pbns.111.04ens.

10 Clayton Whittle et al., *The Environmental Game Design Playbook* (International Game Developers Association, 2022).

11 Whittle et al.

12 Taciano L. Milfont and John Duckitt, "The Environmental Attitudes Inventory: A Valid and Reliable Measure to Assess the Structure of Environmental Attitudes," *Journal of Environmental Psychology* 30, no. 1 (2010): 80–94, https://doi.org/10.1016/j.jenvp.2009.09.001.

13 Dorit Kerret et al., "Two for One: Achieving Both pro-Environmental Behavior and Subjective Well-Being by Implementing Environmental-Hope-Enhancing Programs in Schools," *The Journal of Environmental Education* 51, no. 6 (2020): 434–48.

14 Florian G Kaiser and Urs Fuhrer, "Ecological Behavior's Dependency on Different Forms of Knowledge," *Applied Psychology: An International Review* 52, no. 4 (2003): 598–613.

15 Nita Lauren et al., "You Did, so You Can and You Will: Self-Efficacy as a Mediator of Spillover from Easy to More Difficult pro-Environmental Behaviour," *Journal of Environmental Psychology* 48 (2016): 191–199, https://doi.org/10.1016/j.jenvp.2016.10.004.

16 Whittle et al., *The Environmental Game Design Playbook.*

17 Yuling Zhang et al., "How Important Is Community Participation to Eco-Environmental Conservation in Protected Areas? From the Perspective of Predicting Locals' pro-Environmental Behaviours," *Science of The Total Environment* 739 (October 2020): 139889, https://doi.org/10.1016/j.scitotenv.2020.139889.

18 Jugert et al., "Collective efficacy increases pro-environmental intentions through increasing self-efficacy," *Journal of Environmental Psychology*, 48 (2016): 12–23.

19 Jean Lave, "Situating Learning in Communities of Practice," in *Perspective on Socially Shared Cognition*, ed. L.B. Resnick, 2nd ed. (Boston, MA: APA, 1991).

20 James Paul Gee and Elisabeth Hayes, "Nurturing Affinity Spaces and Game-Based Learning," in *Games, Learning, and Society: Learning and Meaning in the Digital Age*, eds. Constance Steinkuehler, Kurt Squire, and Sasha A Barab, 1st ed. (New York: Cambridge University Press, 2012), 129–153.

21 Hector Postigo, "Modding to the Big Leagues: Exploring the Space between Modders and the Game Industry," *First Monday* 15, no. 5 (2010): 1–14.

22 Blanche Verlie, *Learning to Live with Climate Change: From Anxiety to Transformation*, *Learning to Live with Climate Change: From Anxiety to Transformation*, 2021, https://doi.org/10.4324/9780367441265.

23 Gee and Hayes, "Nurturing Affinity Spaces and Game-Based Learning."

Critical Game Design Paradigms as Tools for Environmental Action

20

Hanna Wirman

ABSTRACT

This chapter establishes critical game design together with speculative design and art games as viable avenues for sustainability action and activism. When applied to game making, these approaches aim at creating play experiences that foster reflection either toward the game form and game development or toward society at large. This chapter investigates examples and techniques of critical/speculative design, art games, and game activism offering inspiration for addressing complex topics with no right answers. It explains how ambiguity and alienation facilitate subjective interpretations and meaning-making and draws on previous similar practices in theatre and film.

KEY TAKEAWAYS

- Understanding of critical game design together with speculative design and art games as viable avenues for sustainability action and activism
- Examples of how critical and speculative design have been used in games to create meaning and foster a reflective position of the player
- Inspiration, guidance, and motivation for making games about and for sustainability topics through "critical game design"

DOI: 10.1201/9781003512400-23

INTRODUCTION: FROM EDUCATION TO REFLECTION

The intersection between digital games and sustainability is perhaps most typically seen in games that aim at *teaching* about topics related to sustainability. Playing these games is expected to lead to behavior changes that manifest themselves in players' everyday life decisions. Such "serious games"[1] have an aim beyond or in addition to entertainment and have gained more culturally and discipline-specific names from "functional games," "applied games," "educational games," "games for change," "social impact games," and "edutainment," to, in some cases, "gamification." Games in this very broad category address sustainability from a range of perspectives. "Recycling games," such as *Recycle Roundup* by National Geographic,[2] might be the most widespread type of sustainability-related games, having a straightforward, focused goal with direct real-life applications for the player. Simulations of circular economies[3] and sustainable forestry,[4] meanwhile, serve as examples of games' unique strength to teach about complex systems through players' interaction with simulations of those systems.[5,6] To think of this complexity in relation to a topic outside of sustainability, consider the *Civilization* game series[7] that efficiently demonstrate how scientific progress is based upon prioritization of time-consuming research initiatives that lead to strategic advantages only when the timing is right in the light of other nations' respective decisions. These games, despite not being intended as "educational" have found their way to classrooms in primary education.[8] Similarly, a multiplayer cooking game like *Overcooked*[9] can introduce some of the team-work aspects and serving culture of a professional kitchen regardless of how unrealistic the actual steps of food preparation of the game might be. Parent and expert reviews alike propose that its players learning valuable skills like cooperation and preparedness for new situations through playing this kitchen simulation game.[10] Often designed to function as part of a formal curriculum, serious games have specific learning goals and are, when rigorously developed, formally assessed regarding how well the learning goals are met.

This chapter looks elsewhere to identify ways of making sustainability-informed games that, instead, rely on alienation, ambiguity, surprise, and subjective interpretation. As such, they leave (or better, make) space for reflection and rely on the curious mind of the player. Recognizing the potential of games to foster social criticism, Gonzalo Frasca quoted Brazilian theater practitioner Augusto Boal while proposing that "it is more important to achieve a good debate than a good solution."[11] It is this kind of relatable yet not immediately applicable, reflective yet not clearly targeted approach that is—so far—rarely seen in games that address sustainability. And it is the aim of this chapter to provide tools for envisioning what such games could look like and to, hopefully, inspire to create them too.

CRITICAL AND SPECULATIVE DESIGN

In the traditions of "critical design" and "speculative design," we can identify alter-natives that add to the repertoire of addressing sustainability in and through games.

These traditions emphasize designer's self-expression and personal agenda together with intended environmental or societal impact over measured learning outcomes. Paradoxically, such games provide a lot of freedom for the players to form their own opinions and attitudes as the game experiences come to build on people's "real world" values and contexts. Consequently, the relationship between designers and players often becomes less rigid and the designer's control over a specific player's experience is reduced. This does not mean that the designer would not be interested in creating impact, influencing the player, or simply having decodable "meaning" in the game but that the designer is rather creating situations for many kinds of interpretations and reflections to emerge instead of leading the player toward specific ones. This assumes a designer who is indeed willing to allow such freedom to a player.

To foster reflection and personal stance, critical games often create, if not rely on, the creation of uniquely intimate play experiences through techniques such as "user-unfriendliness" and "alienation." The core strength of such approaches lies in making the player aware of themselves as players often causing the player to reflect on meanings, values, and ethics that are present in a game or in its play. The aim of this chapter is to identify these and other similar techniques and to underline the underlying "alternative" potential in games that lean toward game activism, activist art, and protest art. This chapter therefore situates sustainability activism done in and through games within the wider traditions of game activism and even art games. Here likeness to art games builds on the "meaning" the player is assumed to gather through play:

> A traditional view of videogames, held by most game developers, is that games do not 'mean' anything [...] art games, however, embedded a point of view through the construction of systemic representation of an idea that produced meaning through a player's active participation.[12]

Critical design in its original context is a narrow branch of, typically, product, and industrial design that encourages users to reconsider and challenge their perspectives of the world. Built on *critical theory*, critical design recognizes and reveals power imbalances, ideologies, and dominant discourses. It aims at igniting social change through reflection as it looks underneath everyday life structures and reveals hidden assumptions. "Artists and activists tend to be the ones who uncover such realities experientially, sometimes by playfully making work that comments on technology itself."[13] Mary Flanagan's work, (*giantJoystick*)[14] is an oft-mentioned critical design (or a piece of game art) that questions the solitary interfaces of videogames—a 3-meters-tall joystick that not only requires collaboration to work but also brings that collaboration to a public space. Pippin Barr's *It is as if you were doing work*,[15] meanwhile, provides a nonsensical but familiar Windows desktop with different software tools to interact with, yet nothing constructive is available to do. The gameplay is filled with seemingly useless but game-progressing and career-advancing clicking of boxes and writing of memos decorated with meaningless stock photos. For Barr, the game is a way to explore the future of work, and for a player, an enthralling and obscure experience of actions that result in feedback that, for a long time, makes no sense at all. The two examples illustrate how critical design in games can take many forms—here, of a hardware modification that changes an existing game and provides self-referential commentary, and of a standalone game that refers to a societal issue while also breaking widely assumed gameplay conventions and challenging the player to reflect on the meaning of this disruption.

When considering critical design in relation to digital games, Lindsay Grace introduces the concept "critical games" and divides them into (1) those that critically engage with topics outside of game development practices and use games as instruments for this purpose and (2) those that self-reflectively comment on the format of videogames themselves—offer "critique of the medium."[16] The two forms, "social critique" and "mechanical critique" are therefore distinct in their perspectives where "social critique looks outward from games toward the society and culture in which they exist" and "mechanical critique looks inward at games from the perspective of game makers or players" (Ibid., 5). While the first category is typically what serious games do, the second fits more authentically to the framework of critical design as known from outside of the game design context. The latter stems from Bardzell and Bardzell's notion of critical design that "foregrounds the ethics of design practice, reveals potentially hidden agendas and values, and explores alternative design values."[17] Similarly, Cecez-Kecmanovic[18] discusses the role of critical theory in Information Systems being to

> reveal distorted consciousness and hidden forms of domination and oppression achieved through or assisted by the use of information systems, the theory also derives its validity from its role in informing and actively engaging in the transformation of IS practice.[19]

Similarly, critical design in the context of games is about critical game creators working on addressing problems related to games and gaming through the games they create. This recursive and self-reflective approach is what Lindsay Grace labels as "mechanical critique." Building on Sherry Turkle's[20] idea about videogames' potential in facilitating "a more sophisticated social criticism" that "would try to use simulation as a means of consciousness-raising," Gonzalo Frasca writes that "Turkle envisions a simulation that would foster its own dissection by letting players to constantly challenge its own rules."[21]

Interestingly, however, it seems that many critical games that have gained notable attention and been recognized for their impact—just like the two examples mentioned before—engage simultaneously with social and mechanical critique. They refer to game (design) cultural or societal issues while also putting the player into a position where their learned knowledge about how games function does not quite suffice in providing the means to succeed in the game. In both cases, there is something about the game system that does not work as expected that makes the player think more carefully about potential and preferred moves—makes them aware of their own actions, their very playerhood.

AMBIGUITY, ALIENATION, AND SELF-REFLECTION

The global ecosystem and human influence on it constitute a "wicked problem" and arguably a "system" that even complex game simulations fail to fully encompass. Game design theorist Miguel Sicart has suggested that designers can foster "ethical agency" in players by introducing wicked problems in games: "Game designers seeking to make their players engage their moral values should create wicked problems in their approach to narrative and gameworld design."[22] In sustainability-related games, these spaces for ethical reflection hence appear almost automatically as they always already deal with wicked problems. Indeed, critical and speculative design invites the player to take the position of an active participant in relation to sustainability but does not give ready-made solutions to how it should be tackled. The designer's intent is on the level of facilitation instead of a specific shift in ways of thinking. It seems reasonable to suggest that the more complicated the issue and challenge at hand, the more important it is to create space for open-minded creativity. After all, complex problems don't have singular, straightforward solutions.

(Critical) game designer and theorist Gonzalo Frasca wrote over 20 years ago that

> the goal of these games is not to find appropriate solutions, but rather serve to trigger of discussions [...] games would work as metonyms that could guide discussions and serve to explore alternative ways of dealing with real life issues.[23]

This idea is fully in line with critical theory's impetus to reveal hidden ideologies to further scrutiny. Frasca's own game *September 12th*[24] provides the player with a possibility to launch missiles into a Middle Eastern town only to make the player feel as if the only appropriate action is to stop doing anything at all as violence in the game encourages more violence and leads to no good at all. The game utilizes its form to highlight a perspective into a specific topic yet leaves space for the player to continue whichever way they may wish, providing no "correct" answers or preferred, let alone quantifiable outcomes. At the same time, the player's position available seems morally gray[25] and likely causes the player to question their acts and values from within this position. The position alienates them from being a player.

Brought to the context of games and sustainability, imagine a game that reminds the player of every kWh of used electricity for playing the game or of the developer air miles that contributed to the creation of the gameplay experience. Or consider a game that only allows vegan players as its users or only functions on devices older than five years. Like numerous examples of realist and modernist theater (and similarly to many popular commercial games these days, too), such games would "break the fourth wall" by speaking directly to the audience, the player. This alienation technique utilized by playwrights like Henrik Ibsen and Bertolt Brecht allows connecting gameplay to related societal challenges on the spot without delay or proxies. "The most important aspect of the alienation is that it draws the players' attention away from the conventions of the medium, and towards a different message or theme."[26] Sicart has suggested that this condition is critical for "ethical gameplay" to take place. Ethical gameplay

> happens as a pause in the fluidity of play—a caesura that forces players to evaluate their behaviors in light of ethical thinking, rather than ludic strategic thinking. Ethical gameplay happens when the game affords a different type of thinking and acting, so that the player as ethical agent is invoked.[27]

Critical game design creates these pauses through alienation and ambiguity that lead to reflection. As such, critical design necessitates acknowledging the player as a person situated in a complex world yet able to think independently and make one's own decisions. Critical game design appears as an attitude and as an ethical approach that builds on more trust in the player than perhaps any other branch of game design. These designs only "work" should the player be able to not only "get them" but meaningfully reflect on their points based on personal life and a wider societal context.

While "talking to the player" (or maybe "with the player"?), critical games often go against player expectations, reduce agency, or create a twist that causes the player to rethink and relearn after an implicit assumption has been revealed.[28] In this regard, critical game design is not far from what is considered "good design" as it pushes the boundaries of the medium. In the award-winning game *Getting Over It with Bennett Foddy*,[29] the player controls a person sitting in a cauldron and swinging a hammer to climb a large mountain. While the controls of the game are extremely difficult to master, a single mistake can cause the player to lose much of their hard-earned progress, making the game particularly frustrating. Bennett Foddy, the designer of the game, suggested in an interview that "whenever you see something that disproves a strongly held design orthodoxy it's extremely exciting because it opens up new avenues for exploration." Typically, his work would not be labeled critical design but rather taken as an example of trailblazing game design in general. Foddy's approach is not exactly critical of the conventions of game design, or the world outside, but simply aimed at finding new paths in unexplored territories. This leads to an exciting outcome: the more conventions and types of games are created and culturally accepted, the more material there is for critical designers to build upon.

In this regard, the goals and means of critical game design are also close to those of what some call *art games*. According to Tiffany Holmes "art games tend [to] challenge one's mental focus in that the player needs to maneuver in the game and simultaneously figure out its conceptual message."[30] Jaron Lanier's *Moondust*[31] is often considered the first art game and established as such given its unusual graphics and sound and gameplay that was difficult to relate due to its uniqueness. The game felt different from anything players had ever experienced and was hard to grasp yet mesmerizing. Cao Fei's *Same Old, Brand New*[32] was a series of videogame symbols and logos viewed on the 490-meter-high LED grid permanently attached to the surface of Hong Kong's tallest skyscraper. This type of an art game—or better game art or game-based art—combined old meanings in new ways and showed them in an unconventional location. The work contrasted the retro symbols from games like *Pac-Man*[33] with the latest viewing technology on a massive skyscraper wall.

John Sharp suggests that art games embed "a point of view through the construction of systematic presentation of an idea that [produces] meaning through a player's active participation."[34] For him, "the play of an artgame [is] intended to have some social, intellectual, moral, or humanistic impact on the player."[35] While similar to many definitions of serious games, art game articulations nevertheless accept the notion of ambiguity and uncertainty in terms of what a game aims at teaching around. While this may not apply to *Moondust*, it certainly applies to many critical games.

USER-UNFRIENDLINESS AND UNCOMFORTABLE ROLES

The ambiguity and alienation of critical games coexist with inefficiency and unnecessary difficulty. Critical design makes consumption less efficient and avoids easiness (cf. Godard's "counter cinema"). This is what leads the user to pause, to reflect, and to challenge and revisit their previous (ideologically dominant) values and assumptions. Anthony Dunne and Fiona Raby, designers who coined the term critical design alongside their own industrial practice, presented it in the opposite end of the spectrum from "affirmative design" that "conforms to cultural, social, technical, and economic expectation."[36] According to them, user-friendliness conceals the fact that all design is always ideological.[37] Critical design, then, offers an alternative to conventional design paradigms through the implementation of "user-unfriendliness." Utilizing user-unfriendliness to force a user to think virtually refers to the same idea as the technique of alienation. These two merely come from different design/research traditions. In both traditions, the user/player is made aware of their role as a player and invited to arrive at their own conclusions around a topic based on the insights and prompts provided by a game.

In Paolo Pedercini's games, published on his website Molleindustria, user-unfriendliness builds on placing the player in a position that is ethically questionable and generally uncomfortable. There is clear progression in Pedercini's games and a conclusion typically, too, but they constantly call for reflection in terms of what would be the right course of action and how atrocious acts one is made to execute. Among others, *Operation: PedoPriest*,[38] which was later removed from Pedercini's website, is a commentary over the paedophilia scandal in the Catholic church and makes the player assume the role of a sex predator. In *Phone Story*,[39] the player explores the dark side of smartphone manufacturing from inhumane conditions of mineral mining all the way to the challenges of dealing with e-waste. Ip Yuk-Yiu's *To Call a Horse a Deer*,[40] which plays with a Chinese saying, makes the player complicit to systematic word-bending and builds on the same experience of uncomfortable role-taking as Pedercini's games.

This characteristic of critical and speculative design also concerns ethics. A critical game, for example, does not always aim at being able to teach the "right" or the "wrong" answer or even claim to be able to suggest possible ways to solve a problem. Instead, it may highlight the problem itself through a contradiction that emphasizes that it should be addressed. In Sicart's terms, they allow "ethical agency."[41] This approach assumes the player as an autonomous and responsible person. Drawing on educational sciences, we may argue that the critical reflection of previous knowledge brought about in critical design is what drives action:

> The primary process by which humans experience emancipation is self-reflection. This is because self-reflective individuals consider the meaning and consequences of their actions. Additionally, self-reflection refocuses responsibility back toward the individual and away from the authority of external experts.
>
> *(Habermas, 1971)*[42]

Reflection through a game is essentially about creating one's *own* thoughts about sustainability based on game experiences not about adopting someone else's views.

The tradition of *speculative design* is even less concrete in its form. Speculative designs are typically mere blueprints or concepts, largely because their goal is to stretch as far into the realms of futurism and utopia/dystopia as needed. This approach to design

> emphasizes inquiry, experimentation and expression, over usability, usefulness or desirability. A particular characteristic of speculative design is that it tends to be future-oriented. This should not be mistaken as being futuristic in a fantasy-like sense,

suggesting that it is 'unreal' and therefore dismissible. Rather, because it is not driven by market imperatives nor by engineering dictates of functionalism, speculative design has the opportunity to imagine and explore possibilities, without the necessity of delivering of actionable plans towards those possibilities. It is thus exploratory and suggestive of what might be.[43]

A critical game that relies on the notion of speculation could, for example, suggest a world where play would all take place using physical exercise stations that self-produce electricity without a need for an external power source. This idea would be presented in a relevant context as a text or a drawing, perhaps an animation, for example. Or it could take the form of a game, too.

CONCLUSIONS

This chapter has identified critical game design, together with speculative design and art games, as viable avenues for sustainability action and activism. I have suggested that critical design is often indistinguishable from art games that, too, critically engage with society, sometimes framed in the context of activist art (cf. Chapter 23 by Jennifer Estaris in this volume). If games and sustainability are too complex a topic to address in commercial games, critical design, and related paradigms boldly tackle issues one at a time, breaking a complex whole into bite-sized approaches that run the risk of oversimplification yet allow pointed commentary without right answers.

Educational psychologist Jerome Bruner has postulated that "much of the process of education consists of being able to distance oneself in some way from what one knows by being able to reflect on one's own knowledge."[44] Bruner's ideas serve as foundational principles in modern educational thought, emphasizing ambiguity, dialogue, and subjectivity. Critical games share these qualities: challenging conventional design and promoting independent thinking. They resist offering definitive answers, instead fostering reflection through alienation, user-unfriendliness, and ambiguity. Designers of such games may lean toward specific interpretations and values but allow players to negotiate and explore alternatives. Bruner further argued that

> it follows from what I have said that the language of education, if it is to be an invitation to reflection and culture creating, cannot be the so-called uncontaminated language of fact and 'objectivity.' It must express stance and must invite counter-stance and in the process leave place for reflection, for metacognition. It is this that permits one to reach higher ground, this process of objectifying in language or image what one has thought and then turning around on it and reconsidering it.[45]

This approach aligns with the ethos of critical games, which prioritize exploration over certainty and embrace the transformative potential of ambiguity.

NOTES

1   Abt, Clark C. 1970. Serious Games. University Press of America.
2   National Geographic. N.d. Recycle Roundup. https://kids.nationalgeographic.com/games/action-adventure/article/recycle-roundup-new (accessed 14 December 2024).
3   E.g. Loop Games. 2015. In the Loop. Board game. https://intheloopgame.com/circular-economy-introduction/ (accessed 14 December 2024).
4   E.g. LVM, University of Latvia and University of Agriculture of Latvia. Mežotājs 2022. https://mezotajs.lv/ (accessed 14 December 2024).
5   E.g. Gee, James Paul. 2006. "Are Video Games Good for Learning?" *Nordic Journal of Digital Literacy*, 1(3): 172–183. https://doi.org/10.18261/ISSN1891-943X-2006-03-02; Gee, James Paul. 2013. The Anti-Education Era: Creating Smarter Students through Digital learning. Palgrave/Macmillan.
6   Such approaches and games have been addressed in Chapter 17 by Anja Venter and Sam Alfred and Chapter 15 by Stefan Werning in this volume, for example.
7   E.g. Firaxis Games. 2001. Civilization III. Infogrames Interactive.
8   King, Matt. 2021. "The Possibilities and Problems of Sid Meier's Civilization in History Classrooms." *The History Teacher*, 54 (3): 539–567. JSTOR, https://www.jstor.org/stable/27058712. Accessed 4 Dec. 2024.; Lee, John K. and Probert, Jeffrey. 2010. "Civilization III and whole-class play in high school social studies." *Journal of Social Studies Research*, 34 (1): 1–28.
9   Ghost Town Games. 2016. Overcooked. Team17.
10 Learning Corner. https://learningcorner.co/activity/137022 (accessed 14 December 2024); Learning Works for Kids. https://learningworksforkids.com/playbooks/overcooked-2/ (accessed 14 December 2024).
11 Frasca, Gonzalo. 2001b. Videogames of the oppressed: Videogames as a Means for Critical Thinking and Debate. MA Thesis, Georgia Institute of Technology.
12 Sharp, John. 2015. *Works of Game: On the Aesthetics of Games and Art*. MIT Press, 53–54.
13 Flanagan, Mary. 2009. Critical Play: Radical Game Design. MIT Press, 225.
14 Flanagan, Mary. 2006. [giantJoystick].
15 Barr, Pippin. 2017. It is as if you were doing work.
16 Grace, Lindsay. 2014. "Critical Games: Critical Design in Independent Games." DiGRA 2014 Proceedings, 1.
17 Bardzell, Jeffrey and Bardzell, Shaowen. 2013. "What is "critical" about critical design?" CHI '13: Proceedings of the SIGCHI Conference on Human Factors in Computing Systems, 3297–3306. April 2013. https://doi.org/10.1145/2470654.2466451.
18 Cecez-Kecmanovic, Dubravka. 2005. "Basic assumptions of the critical research perspectives in information systems." In Debra Howcroft, Eileen Moore Trauth (eds.): Handbook of Critical Information Systems Research: Theory and Application, 19–46, Edward Elgar Publishing.
19 Cecez-Kecmanovic 2005, 40.
20 Turkle, Sherry. 1995. Life on the Screen. Simon & Schuster.
21 Frasca, 2001b, 58.
22 Sicart, Miguel Angel. 2013a. *Beyond Choices: The Design of Ethical Gameplay*. MIT Press, 35.
23 Frasca, Gonzalo. 2001a. "Videogames of the Oppressed: Critical Thinking, Education, Tolerance and other Trivial Issues." In Noah Wardrip-Fruin and Pat Harrigan (eds.): *First person: New Media as Story, Performance, and Game*. MIT Press, 85.

24 NewsGaming.com. 2003. September 12th: A Toy World. http://www.newsgaming.com/games/index12.htm (accessed 14 March 2025).
25 Aeschbach, Lena Fanya, Opwis, Klaus and Brühlmann, Florian. 2022. "Breaking Immersion: A Theoretical Framework of Alienated Play to Facilitate Critical Reflection on Interactive Media." *Frontiers in Virtual Reality*, 3: 846490. https://doi.org/10.3389/frvir.2022.846490.
26 Dunne, Daniel. 2014. "Brechtian Alienation Theory in Videogames." *Press Start*, 1(1): 80–99.
27 Sicart, Miguel. 2013b. "Moral Dilemmas in Computer Games." *Design Issues*, 29 (3): 28–37. https://doi.org/10.1162/DESI_a_00219.
28 Wilson and Sicart's (2010) definition 'abusive game design' focuses specifically on the personal dialogue between the designer and the player through such games. (Wilson, D. and Sicart (Vila), M. A. 2010. "Now It's Personal: On Abusive Game Design." In Future Play Conference Proceedings (pp. 64). Association for Computing Machinery.)
29 Foddy, Bennett. 2017. Getting Over It with Bennett Foddy.
30 Holmes, Tiffany. 2003. "Arcade Classics Spawn Art? Current Trends in the Art Game Genre." *Proceedings of the 5th International Digital Arts and Cultures Conference*, Melbourne, Australia, 46.
31 Lanier, Jaron. 1983. Moondust. Creative Software.
32 Cao, Fei. 2014. Same Old, Brand New.
33 Namco. 1980. Pac-Man. Namco.
34 Sharp 2015, 53.
35 Sharp 2015, 54.
36 Dunne, Anthony and Raby, Fiona. 2001. Design Noir: The Secret Life of Electronic Objects. Birkhäuser Press; cf. Flanagan, Mary and Nissenbaum, Helen. 2014. Values at Play in Digital Games. MIT Press.
37 Dunne, Anthony. 1997. Herzian Tales. PhD thesis. RCA.
38 Molleindustria. 2007. Operation: Pedopriest. https://www.molleindustria.org/en/operation-pedopriest/ (accessed 14 March 2025).
39 Molleindustria. 2011. Phone Story. https://www.phonestory.org/ (accessed 14 March 2025).
40 Ip, Yuk-Yiu. 2018. To Call a Horse a Deer. https://www.ipyukyiu.com/tocall (accessed 14 March 2025).
41 Sicart 2013a.
42 Denton, David. 2011. "Reflection and Learning: Characteristics, Obstacles, and Implications." *Educational Philosophy and Theory*, 43(8): 838–852. https://doi.org/10.1111/j.1469-5812.2009.00600.x.
43 DiSalvo, Carl and Lukens, Jonathan. 2009. "Towards a critical technological fluency: The confluence of speculative design and community technology programs." In *Proceedings of the Digital and Arts and Culture Conference*, 2009: After media: Embodiment and context, n.p. http://escholarship.org/uc/item/7jz308ws (accessed 14 December 2024).
44 Bruner, Jerome. 1986. *Actual Minds, Possible Worlds*. Harvard University Press, 127.
45 Bruner 1986.

SECTION 4

From Theory to Action

Section 4
From Theory to Action

21

Patrick Prax

The final section of this book is about bringing all of our topics together: the work for material systemic change, the consideration of the production context of games, and their design, message, and strategy. In doing so, this section aims to close the circle that we opened at the beginning of this book with the enormous challenge of meaningful sustainability work. But it also adds something that has so far been missing: the perspectives of the human beings doing all this—the perspective of you, the game maker.

The section starts out with Ossian Nordgre and Kristofer Vaske who investigate games and game-making as resistance to a destructive status quo. This chapter has a split focus here, not only analyzing four games, Chicago '68, Which Side?, Disarm the Base, and Zone to Defend, but also connecting to the personal experiences and perspectives from the game makers who made these games as a kind of resistance. This connection of game-making to personal, lived experiences about resistance and activism connects the strategies of their resistance movements to game-making as a strategy for resistance more broadly. Experiencing agency together and collectively keeping histories of resistance alive is a crucial step towards meaningful change and fighting back.

Jennifer Estaris's chapter also lives at the intersection of personal political engagement and activism through making games. Jennifer is an example of how it is not only possible to combine activism and game-making, but that activism can even inform and inspire one's creative work and help create games that resonate with players on a sincere, human level. She also shows what the path of a designer coming to this kind of game creation can look like, and that there is a community of people already doing this work who can help with building the important connections needed to thrive as a game maker with a mission.

This leads to the final and maybe most personal chapter in this book, where Hexe Fey kindly shares her personal strategies and reflections on surviving as a game maker

DOI: 10.1201/9781003512400-25

with a focus on sustainability and human connection. Hexe draws heavily on her own experiences, in connection with indigenous cultural ontologies and political research, and, of course, also from this perspective, stresses the importance of creating a community for support and being there for each other.

This leaves Patrick Prax, Clayton Whittle, and Trevin York with their conclusion to this book.

Rendering Resistance

22

Making Defiant Games

Ossian Nordgren and Kristofer Vaske

ABSTRACT

Slow, incremental, and minor actions toward mitigating climate change dominate policy discourse and much of "green" game design. Despite the extreme urgency of the situation, industries and systems destroying our conditions of life are booming. This chapter argues that we need to go beyond just learning about sustainable living, we need to actively resist. Based on interviews with games makers and readings of their games, this chapter explores how resistance games can be a vessel for game makers and players to imagine and engage with climate struggle and green games in ways leading to meaningful material and social change.

KEY TAKEAWAYS

- An actionable understanding for game makers to connect and incorporate themes and mechanics of historical and contemporary resistance movements in their designs
- Games as a way for developers and players to identify and engage with resistance struggles working toward human rights and climate justice
- Powerful examples of *resistance games* that can inspire future exploration and innovation within the genre.

DOI: 10.1201/9781003512400-26

INTRODUCTION: GAMES AS HORIZONS OF RESISTANCE

If the design and industry of games, as suggested by the title of this collected volume, *to need change* and *move toward sustainability*, we as developers, academics, and otherwise concerned people must seriously engage with how this change is expected to come about. The impending climate crisis and the ongoing 6th planetary extinction event expose the severe urgency of this historical moment.[1] We have been positioned at a precipice where meaningful action toward reducing humanity's planetary impacts has become an inescapable obligation.[2] This chapter invites game makers to use their craft to foster resistance against the systems driving climate change, drawing on interviews with developers already doing this work. By exploring games that directly confront and resist the very structures, powers, and institutions at the heart of climate change, we can at least *move against* that which is hurting us—in a way, removing the knife before tending to the wound.

Mainstream climate solutions often emphasize gradual reforms like carbon taxes, green technologies, and individual behavior change, typically promoted through institutional frameworks and voluntary initiatives.[3] Many "green games" reflect this incremental approach.[4] We problematize the sufficiency of such strategies and ask if they are merely reinforcing the status quo. Instead, we propose *Resistance games* as an alternative. Resistance has long been a common theme in games, often depicted through familiar narratives of rebellion against oppressive regimes, yet games that treat resistance as a serious, real-world political tool are less common. Drawing from historical and contemporary struggles, *Resistance games* advocates for designs that expand the boundaries of political possibility.[5] These games challenge incrementalism by centering resistance and envision radical transformations beyond the limitations of mainstream environmental discourse and politics.

Our interviews reveal the diverse ways resistance has been imagined and implemented in both tabletop and digital games, with a focus on the former. By examining game mechanics and design decisions, we show how subtle changes can lead to profound shifts toward games of resistance.

This chapter analyzes four key games—*Chicago '68,*[6] *Which Side?,*[7] *Disarm the Base,*[8] and *Zone to Defend*[9]—highlighting how each exemplifies different aspects of resistance, from protest and labor struggle to sabotage and direct action. These examples illustrate how game design can creatively has incorporate and support real-world movements for radical change. Finally, we provide a few actionable takeaways for game makers attracted by the potential of *resistance games.*

Mainstream Climate Action and Green Games

Mainstream climate mitigation efforts typically follow recognizable patterns: emissions budgets, taxation, circular economies, green technologies, and voluntary incentives for decarbonization.[10] Other mitigation strategies focus on behavioral change of individuals, promoting recycling, sustainable consumption, and low-carbon transportation, or facilitating a shift in our relationship to the more-than-human world.[11] "Green games" often align with these incrementalist frameworks and the fundamental idea that by fostering awareness, gradual change, or recomposing within existing structures, climate breakdown can be averted[12] (see Lopez and De Masi et al. in this volume for more).

While these approaches may have some merit, we must ask whether they limit possibilities by adhering to a narrow idea of climate mitigation, often shaped by those with vested interests.[13] Additionally, if we are to take the increasingly desperate calls of climate scientists seriously, we must ask if the time for incremental, slow, or marginal mitigation has passed. Green games with ecological themes, recycling mechanics, or nature preservation storylines surely have a place in the collective struggle, but they should not have exclusive claim to what gamified climate action can be.[14] *Resistance games*, in contrast, propose mitigation based on empirical understandings of historical radical societal transformation, the frightening depth of societal carbon dependency, and the repeated unwillingness of powerful actors to willfully forgo their influence and resources.[15]

Resistance

> Recognizing the direness of the situation, it is high time for the movement to more decisively shift from protest to resistance: 'Protest is when I say I don't like this. Resistance is when I put an end to what I don't like.[16]

As a thematic device or backgrounding setting, resistance has seen heavy utilization in cultural works, not least in games. Be it in Star Wars,[17] Mirrors Edge[18] or Final Fantasy XIII[19] playing the part of an outnumbered and underfunded resistance force set against an unjust or evil despot has obvious merit as a plot device. The content, however, has generally not been seen as related to real-world issues and parallels to actual politics or historical conflicts have often been obfuscated[20]

Spaces, movements, and acts of resistance are regularly present in moments of historical and contemporary social significance. Taking forceful shapes in slave uprisings and guerilla opposition in occupied wartime nations and non-violent silhouettes in civil rights movements and labor strikes, resistance comes to us in diverse compositions. These instances come together in their organized opposition to and direct actions against that which is occupying, subjugating, invading or harming them and their oppressed compatriots.[21]

We can understand resistance as both organized and spontaneous acts by collectives and individuals that actively oppose oppression or unjust brutal violence. Resistance is commonly associated with WW2-occupied Paris or the Haitian Revolution. But ties to ecological circumstances are easy to make if we consider the totalizing violence and suffering of climate breakdown on top of the local ferocity of natural resource extraction and ecosystem degradation.[22] A through line can thus be drawn between workers striking in the face of horrendous working conditions in the 1900s New York textile industry (*Which Side?*) and activists blocking the development of an international airport today (*Zone to Defend*), both parties to a greater struggle spanning time and space.

The word resistance can mean many things for different people.[23] For the purposes of this chapter, we have drawn conceptual inspiration from the writings of human ecologist Andreas Malm. Resistance for Malm,[24] as alluded to in the quote at the top of this section, goes beyond mere refusal or political protest. Instead, he figuratively instructs the readers of his most influential work on "How to blow up a pipeline."[25] While Malm is concerned neither with game design nor its industry, his writing provides concrete political parchment to map our designs upon.

We base our conviction of the power of resistance both on its historical merit and its contemporary effectiveness. Resistance by a small radical flank has played major, but often underappreciated parts, in most, if not all, majorly successful social movements, for example during the struggles of slave emancipation in the US or the Indian Freedom Struggle.[26] *Place-based-resistance* has been shown as an effective approach in halting and postponing fossil fuel or other ecologically damaging infrastructural developments, 25% of projects which encounter such resistance were majorly disrupted.[27]

RESISTANCE GAMES—A RESISTING LUDOLOGY

Our interviews have given insight into ways in which forms of resistance can and have been imagined and implemented in games and how these relate to one another. Our total catalog of *resistance games* is split between table-top and digital games, but our interview sample leans heavier on the former.

This chapter below is structured around four main games. We begin by looking into social movements through (1) *Chicago '68*, a political war game which "pits revolutionary spectacle against civil order at the Democratic National Convention riots of 1968."[28] Next, we describe how *Which Side?* (2) explores of labor movement resistance in "a historic and cooperative survival game that simulates the workers' struggles"[29] from Yeast Games. We continue delving into resistance through sabotage and blockades with the former exemplified by players being tasked to "work together to find war planes, disarm them and escape before it's too late!"[30] in *Disarm the base* (3) and the latter illustrated in *Zone to Defend* and its occupants "fight against a destructive and harmful airport project."[31] The examples are selected for their indicative portrait of key tenets of resistance, creativity in incorporating these within a game as well as their boldness in taking controversial and imperative stances (see McBay and Holowka in this volume for more about games connected to social movements).

Chicago '68—Conflict as Spectacle

Set during the tense days of the 1968 Democratic National Convention in Chicago, Chicago '68 reimagines war gaming through underexplored perspectives and novel tactical elements.[32] Players navigate the mass-protest and try to influence the party nomination proceedings inside *The International Amphitheater*. While clashes between protestors and police are built into the game, actions extend beyond combat, pitting "revolutionary spectacle against civil order."[33] The game replaces traditional combat with a conflict system that considers both violent and moral actions. Media coverage and public perception dictate the decision space and available actions. The establishment faces internal challenges like capital limits, while demonstrators deal with internal dissent. The shifting decision space, resource access, street actions, and use of force aim to balance gameplay and approximate the historical protests of Chicago in 1968.

In making a game depicting mass social movements, a core negotiation comes with *player perspective*. In classical war or social movement games, "you play as [its] nebulous spirit," *Chicago '68* Developer Joe Dewhurst noted. Games of strategy tend to place the player in an ominous position of power, acting as military general or great planner Joe continued. Players are commonly installed in a world built on fictive or approximate historical grounds and tasked with defeating the enemy by winning battles and controlling territory. Joe stressed this tension in postulating "What does winning look like in street protest? What tools get you there? Holding ground like armies doesn't make sense...." *Chicago '68* approaches this via *exposure* where *the demonstrators* win by capturing "the hearts and minds of the American public" through their street actions, while *the establishment* wins by quashing the uprising and electing Hubert Humphrey as the Democratic presidential candidate.

Parallel to player perspective the issue of historical simulation was central to lead designer Yoni Goldstein who asked, "What does it mean to represent a riot with millions of points of conflict?". Joe builds on this notion by pointing out that a lack of real time information is a key experience of actual social movements engaged in demonstration and how hard this is to emulate in board games "as the player [inherently] has control over the whole board." *Chicago '68* partly addresses this challenging task using historical reference material (the game being based on a government report containing thousands of eyewitness and police accounts, maps, and reporting) and by recognizing the necessity of contentious historical interpretation. What separates games like these from commercial war games is a deep consideration of simulation. Commercial games often begin with "mechanics with theme added on later," Joe concludes. Making historical social movement games does often, and probably should, work in the opposite direction.

> What is a riot if not a production of historical evidence?
>
> *Yoni Goldstein*

Fully accurate representations are rarely the goal; compromise and simplification are needed. "The game is not a historical simulation, there is no determinism... What we get instead is more like folk song, a narrative way to remember it 50–60 years later," Yoni declared. Games as artifacts narrating history became a recurring theme in our conversations with designers, which we shall return to below.

Simulating an uprising, riot, or mass protest requires careful consideration. Yoni and Joe have taken the events of Chicago '68 seriously. At its core the forms of resistance unfolding in *Chicago '68* and the ways these have been deployed in the game are not exclusive to that context. They are translatable to other historical, contemporary, or fictive struggles aimed at those implicated in the heating of our planet, like mass-protest at the sites of gatherings of fossil fuel executives or locations of natural resource extraction.

Summarized in its rulebook *Chicago '68* is "...a game about the kinetic and political struggle over historical narrative." Game makers interested in groups like Extinction Rebellion, Letzte Generation, or Fridays for Future can draw inspiration from this summary. As much as climate resistance is a kinetic and material struggle to mitigate climate change, it's also a conflict over historical narrative. By building on conventional mechanics *Chicago '68* offers familiarity with unexpected modifications. For example, by implementing street theater and spectacle as alternatives to combat, leaning into a partial player game-state-oversight or adding layers of contention such as the media landscape. A stark awareness of the limits of simulation and perspective together with consideration of history and theme over mechanics shows how to approach the complex dynamics of social movement games.

Which Side?—Agency of the Marginalized

As game developer Alexandre Weiss was finishing his PhD in history, he recognized that his dissertation on historical syndicalism in the US and Mexico would work well reimagined as a game. He combined his interests and skills to create *Which Side?*, a cooperative board game simulating worker struggles in New York during "the height of the fires in the textile industry." The game explores an alternative kind of resistance to *Chicago '68,* one which drew its motivation from working-class survival and labor conditions rather than national politics.

Each player takes the role of a *tenement* with 4–5 inhabitants and, using worker placement, chooses their daily and nightly activities. Some must work or resort to criminal activity to afford rent, while others engage in political organizing. Depending on their choice, the risk of workplace accidents, housing conditions, and the threat of police violence or jail will be adjusted. The game is a constant endeavor of using the scarce resources to enable strikes, sabotage, and agitation to counter the employer's strategy cards. Players then have the opportunity to improve their living conditions and take power back from their oppressors. The game's cards and systems, along with historical detail, make clear the perilous life and struggles of the early 20th-century working class.

Alexandre furthers the points made by Yoni and Joe and describes his goal of telling

> [the] histories of the worker far away from the state ... [thus] giving the workers an alternative history" and "showing the employers as they are; they kill people, they have the cops on their side, they pollute the environment" culminating with the game's ultimate purpose of "changing the perception of history.

Joe Dewhurst claimed "a game like this allows you to understand historical agency, choices in the context ... What a game is doing is something different from other mediums." When designing within real historical contexts and aiming toward fostering resistance, agency trumps historicity or documentary certitude. In contexts of oppression and violence engaging with the agency of the oppressed becomes vital. Understanding that historical choices lay the foundation of our socio-economic reality and ecological circumstance crystalizes the human-made and socially constructed nature of climate change. Our current moment is no exception; therefore learning from those who came before us, took massive personal risk, and acted beyond self-interest in dire circumstances becomes a crucial point of departure.

Games deal with *agency in context* especially well by emplacing players as decision makers, forcing them to take risks and act in situations with limited knowledge of outcomes. Thus, as Alexandre and Joe point out, a well-designed game where agency in historical context is taken seriously can empower us to reconsider how and what we can do to overcome that which subordinates us, be it state forces squashing an uprising, corrupt politicians in bed with textile barons or fossil fuel magnate's green washing attempts.

You don't have to change the history – it's all there, just can just show it

Alexandre Weiss

Alexandre extends the politics of *Which Side?*[34] to the process of manufacturing. With deep reverence for the early 19th-century working classes of New York he developed the core theme of the game, in a similar vein, he decided to fabricate it with current worker relations in mind. Rather than working with suppliers and factories in the global south he sourced 95–97% of material from within 50 km of his residence, used a local printer, and assembled everything by hand with the help of his collaborators and long-standing local LARPing community. Despite this effort being incredibly laborious, time-consuming, and expensive, Alexandre justified it on ethical and ecological grounds. Through avoiding plastic and intercontinental distribution, the direct environmental impacts of the game were greatly reduced. Through "obtaining [their]own means of production" oversight and control over the circumstances of those assembling the game were facilitated. This later justification truly brings together the content and context of *Which Side?* as Alexandre stands by it as an act of defiance and resistance in the face of increasing corporate consolidation of board games production and publication and the ecological and social ramifications of this process.

Revisiting historical movements, such as the '68 demonstrations or the workers movement of early 20th-century New York, with an alternative focus and perspective, through the medium of gameplay, enables the player to emerge and situate themselves in a historical unfolding which offers an opening for reimagining current social relations and forms of politics. Despite the way that industry and systems can seem to have power over people, they are deeply reliant on their cooperation—a cooperation that can be revoked. Illustrating this in a game and presenting it as relatable to struggles of ordinary people can bring attention to the fact that history is rarely changed by individual heroes, but rather by the collective force of people like you and me deciding that we've had enough.

Disarm the Base—Striking Where It Hurts

While Jessica Metheringham was working with parliamentary engagement at *Quakers in Britain* in 2017, a colleague of hers was part of a peace protest direct action trying to break into a military base and disarm war planes, inspiring her to make *Disarm the Base*. She tells us, "They [Quakers] are a religious movement, but they are very protesty[...] They're one of the peace churches." This and a similar action in 1996, both attempted to prevent Britain from exporting war planes to Saudi Arabia and the

Indonesian military respectively. While the 2017 action was stopped before reaching the planes, the activists in 1996 caused 1.5 million pounds worth of damage to a plane before calling security on themselves.

> Welcome to the disarmament movement. Inside this military base, planes are being armed with deadly weapons. Tomorrow morning, they will fly halfway around the world to bomb civilians. Can we stop them?[35]

Nine hangars are placed face down on the game board with warplanes in 4–6 of them. The game is in large part dictated by a deck of cards. It holds both items needed for the activists to complete their quest, like crowbars and bolt cutters, and cards determining the actions of security personnel, like activating flood lights, dispatching patrolling guards or closing gates. If the deck runs out, the sun rises and the activists have failed. Players move freely around the board but as the game goes on more and more security guards and lights activate, increasing the risk of being found and captured before disarming every plane. Depending on available cards, players might be able to enter the control tower and signal for guards to leave a hangar, or resort to making a distraction, achieving the same goal but putting the guards on alert, causing more of them to appear next time. The game highlights the risk of pulling off such direct actions and how easy it can be for things to go wrong.

> We'll need to work quickly, but also be careful. Stay out of the light, stay out of the buildings, and we should be ok. Good luck![36]

Jessica developed the game in collaboration with the *Campaign Against Arms Trade* (CAAT) and the revenue goes toward their cause. She tells us how, despite being about sabotage, this resulted in an added importance of the game otherwise being non-violent, to comply with the strategies and philosophy of the CAAT. "It's very easy to have a game which mechanically draws you towards a fight. [...] I don't think that it necessarily has to be that way, but because we've done them so often, this is what you do," she continues. By tying the game and its development to an existing movement, a target audience is already there to inform the decisions of the designer.

While *Which side?* covers a longer struggle, *Disarm the Base* hones in on a particular action set during a few hours. In the same vein, we can experiment with how games can represent ground level perspectives, and small-scale action instead of attempting to capture a climate movement in its entirety. Although the struggle for mitigating climate change requires resistance on a global scale, each small part will be carried out by a person with a unique perspective, context, sense of place, and motivation, often requiring personal sacrifices. Games have the potential to portray seemingly small events that nonetheless can have great impact.

Though the setting and historical moment differ, both *Which Side?* and *Disarm the Base* spotlight the possibility to literally and figuratively throw a wrench into the machinery of oppressors to demand a world where the lives and rights of people weigh heavier than the profits of business or the arms industry. Enveloped by systems inching ever closer to planetary collapse, one may sometimes feel powerless. When we as game designers search for ways to illustrate possibilities of resistance against the systems responsible for climate breakdown we, just as everyone else, should look for points of weaknesses where ordinary people have opportunities to put pressure

Property does not stand above the earth; there is no technical or natural or divine law that makes it inviolable in this emergency. If states cannot on their own initiative open up the fences, others will have to do it for them.[37]

If the threat to our climate and environment feels intangible and eluding, then what *Disarm the Base* reminds us is that machines and infrastructure responsible for creating human suffering are physical in nature and can be broken.

Zone to Defend—Place-based Resistance

Zone à Défendre[38] or "Zone to defend" (ZAD from its French origins) is an expression referring to "a militant occupation that is intended to physically blockade a development project"[39] The board game carrying the same name depicts one of the most notable examples of this, the ZAD of Notre-Dame-des-Landes (Northwestern France) in which activists blocked and occupied the prospective area for more than ten years eventually stopping the construction of a new international airport. What began as local farmers and inhabitants refusing to leave their land grew into a large movement embodying environmentalist and anti-capitalist ideals. The ZAD's motto "Against the airport and its world" together with the on-site alternative community that took shape, including experiments of autonomous self-sustained living through farming and mutual aid, conjoin to exemplify how direct actions of environmental resistance and wider modes of organizing collective life outside of conventional systems can emerge successful.

In the game, players take the role of the occupying activists, resisting the pervasive threat of police evictions and halting the advance of bulldozers. Winning is contingent on collecting *Struggle Points,* these gathered by generating the *food* and *support* needed to sustain the movement while simultaneously stopping the bulldozers and their police escorts from starting construction of the airport. The players acquire *skill cards* representing the diversity of tactics used in the struggles at the ZAD. These contain building constructions such as farms, cabins, bakeries, or barricades but also efforts connected to the confrontations with police and media. There's clown activism, hacktivism, filming actions, crafting banners, baking cookies, sabotage, setting barricades on fire or even throwing Molotov cocktails—a multitude of creative strategies mirroring the spectacles of the 1968 demonstrators in Chicago.

> Some pursued non-violent direct action while others threw Molotov cocktails. On the ground it confused the cops, one minute they faced old age pensioners singing songs, the next a burning barricade. In the media battle of the story, the state could not capture us in their classic trap, they could not divide us with the 'good' vs 'bad' protester stereotypes[40]

All of these actions contribute to gathering *support, food* or stopping the bulldozers' advance on the game board. *Food* is required to keep all activist pawns active and useful as well as united, with *support* being the currency for completing *projects*, which grant the *Struggle points* needed to win the game. The *project cards* contain different actions connected to the history and struggle on the ZAD such as creating a human chain, organizing a climate action camp, building a bike tractor or setting up a demonstration. Defensive constructions and acquired skills help the players in confronting and delaying the bulldozers advancing on the board, but for every victory, and as the game progresses, the repression of the system increases and ensures that the next bulldozer will have a larger escort.

The game was crowdfunded and produced to support and finance the actual ZAD. Its development was initiated by Sacha Epp,[41] after having lived at the ZAD for over a year and was completed with the efforts of over 30 volunteering testers, artists, and developers. The game documents events that took place in the struggle and mirrors the ZADs resistance and victory against the systems that be. It relates in different ways to all of our previous examples, featuring historical representation, cooperative struggles for subsistence and mobilization as well as authentic examples of diverse techniques and strategies for direct action.

The ZAD of Notre-Dame-Des-Landes is an example of the types of actions and forms of prolonged and wide resistance that succeed in their mission. In 2018, when the airport project was canceled, the ZAD joined the club of similar placed-based resistance movements that achieved their aims. As mentioned, this kind of place-based resistance has shown to be a successful recipe, successfully disrupting a fourth of targeted development projects. For the sake of everyone looking for hope and guidance in this moment in history, this kind of struggle and victory needs to be portrayed more often in all kinds of media but maybe especially in games. Finding these examples and spreading the word by putting them in the hands of players to learn from and immerse themselves in is in itself a goal to pursue.

> The book was originally envisioned as a call to arms for the English-speaking world, [...] Eager to go to press, the book's deadline somehow nevertheless kept on being pushed back... because the people of the ZAD kept winning. It became apparent that we wouldn't be publishing a call to arms for a battle yet to come, but rather something else, a reflection on victories.[42]

CONCLUSION: GAMES AS VESSELS FOR RESISTANCE

Across these case studies—*Chicago '68, Which Side?, Disarm the Base,* and *Zone to Defend*—clear patterns emerge: *Resistance games* encourage us to rethink how we model conflict, agency, activism, and historical agency in game design. These games foreground collective action, moral legitimacy, and the everyday struggles embedded in mounting resistance. They reframe conflict not as a contest of arms and armies, but as a battle over public perception, solidarity, and survival.

An important consideration for game makers is to shift focus from individual heroics to collective agency. Success in these games depends on sustained cooperation, shared risk, and community resilience. They show how decentralized movements can wield real power—and how games can simulate that dynamic by embracing partial control, asymmetric information, and fluid decision spaces that evolve in response to social legitimacy and systemic pushback. The actions of collectives to enact real material impacts, be it through sabotage or mutual aid, are also aptly revealed.

Moreover, these games have built systems that allow players to experience their agency in historical context. Systems that reveal both the oppressive power of structural forces and invite players to engage with counter-histories and alternative narratives, highlighting the experiences of those often marginalized in traditional historical accounts. Whether through the symbolic street theater of *Chicago '68* or the embodied resistance of *Zone to Defend*, these games evoke the diverse realities of resistance, not just its strategic mechanics. Telling these narratives also functions to legitimize the form of resistance occurring within them. Showing the efficacy of prolonged place-based resistance to hinder ecologically harmful developments, for example, can provide tangible examples to inspire others and offer actionable hope.

Crucially, the politics of these games extend beyond play. Designers like *Alexandre Weiss* bring their values into the production process, using local materials, ethical labor, and directing proceeds to real-world movements. The creators of *Zone to Defend* illustrate how *resistance games* can be a part and product of contemporary place-based resistance movements. By integrating the values of the Quakers peace church into *Disarm the Base*, Jessica Metheringham exemplifies how working with existing movements can provide design direction and thematic coherence. These approaches warrant that the means of game creation align with its political messages, deepening the authenticity and commitment of the project.

Taken together, these examples point to a radical reimagining of game design as a tool for fostering resistance. By centering consequential historical and contemporary events, hopeful uncertainty, and the power of ordinary people to effect change, *Resistance games* offer players more than entertainment—they provide a space to rehearse collective struggle and imagine new worlds. For game makers interested in engaging with resistance, these examples offer a compelling blueprint: games as interactive media and platforms for resisting climate breakdown.

TABLE 22.1 Lists a Selection of Resistances Games and Their Associated Creator/Designer

TITLE	CREATOR/DESIGNER
Defenders of the Wild	Outlandish Games
ANTIFA the Game	La Horde
Okupa tu també! (Let's squat!)	Gall Negre
RIOT: Cast the First Stone	No Board Games
Freedom Underground Railroad	Brian Mayer
Bloc by Bloc: Uprising	Outlandish Games
Autonomía Zapatista	Collaborators in Solidarity with the Tour for Life 2021
A Las Barricadas—A Boardgame of Social Conflict	Juan Carlos Cebrián, Nicolás Eskubi
STRIKE!: The Game of Worker Rebellion	TESA Collective
The Fully Automated Board Game	No Board Games
Riot: Civil Unrest	Leonard Menchiari
The Saboteur	Pandemic Studios
Tonight we Riot	Pixel Pushers Union 512
They Came from a Communist Planet	Colestia
Sabotage Will Set Us Free	Machinescreen

APPENDIX 1: *RESISTANCE GAMES* DATASET

In addition to the games mentioned in this chapter, the table below lists some of the titles that we've looked at during the writing process. Though we didn't have enough space to discuss them all in this chapter, we think they might be valuable suggestions for readers interested in exploring the topic more, to save you the time of having to look them up as we did.

NOTES

1 Richardson, Katherine, Will Steffen, Wolfgang Lucht, Jørgen Bendtsen, Sarah E. Cornell, Jonathan F. Donges, Markus Drüke, et al. 2023. "Earth beyond Six of Nine Planetary Boundaries." *Science Advances* 9 (37): eadh2458. https://doi.org/10.1126/sciadv.adh2458.
2 Calvin, Katherine, Dipak Dasgupta, Gerhard Krinner, Aditi Mukherji, Peter W. Thorne, Christopher Trisos, José Romero, et al. 2023. "IPCC, 2023: Climate Change 2023: Synthesis Report. Contribution of Working Groups I, II and III to the Sixth Assessment Report of the Intergovernmental Panel on Climate Change [Core Writing Team, H. Lee and J. Romero (Eds.)]. IPCC, Geneva, Switzerland." *First. Intergovernmental Panel on Climate Change (IPCC).* https://doi.org/10.59327/IPCC/AR6-9789291691647.
3 Hickel, Jason. 2021. *Less Is More.* London, England: Windmill Books. ISBN: 9781786091215.
4 Wu, Jason S., and Joey J. Lee. 2015. "Climate Change Games as Tools for Education and Engagement." *Nature Climate Change* 5 (5): 413–418. https://doi.org/10.1038/nclimate2566.
5 Dyer-Witheford, Nick, and Greig de Peuter. 2021. "Postscript: Gaming While Empire Burns." *Games and Culture* 16 (3): 371–380. https://doi.org/10.1177/1555412020954998.
6 Yoni Goldstein. *Chicago '68.* The Dietz Foundation. Board game. 2024.
7 Alexandre Weiss and Franck "Counit" Catoire. *Which Side?.* Yeast Games. Board game. 2021.
8 Jessica Metheringham. *Disarm the Base.* Dissent Games. Board game. 2019.
9 *Zone à Défendre.* Contrevents. Board game. 2017.
10 Hickel, 2021.
11 Hornborg, Alf. 2001. *The Power of the Machine: Global Inequalities of Economy, Technology, and Environment. Globalization and the Environment.* Walnut Creek, Calif.: AltaMira Press. ISBN: 978-0-7591-0067-1.
12 Fernández Galeote, Daniel, and Juho Hamari. 2021. "Game-Based Climate Change Engagement: Analyzing the Potential of Entertainment and Serious Games." Proc. ACM Hum.-Comput. Interact. 5 (CHI PLAY): 226:1-226:21. https://doi.org/10.1145/3474653.
13 Supran, Geoffrey, and Naomi Oreskes. 2021. "Rhetoric and Frame Analysis of ExxonMobil's Climate Change Communications." *One Earth* 4 (5): 696–719. https://doi.org/10.1016/j.oneear.2021.04.014.
14 Hammar, Emil L., Carolyn Jong, and Joachim Despland-Lichtert. 2023. "Time to Stop Playing: No Game Studies on a Dead Planet." *Eludamos: Journal for Computer Game Culture* 14 (1): 31–54. https://doi.org/10.7557/23.7109.
15 Malm, Andreas, and Wim Carton. 2024. *Overshoot: How the World Surrendered to Climate Breakdown.* London New York: Verso. ISBN: 978-1-80429-398-0.
16 Malm 2021: 70.

17 George Lucas. Star Wars. Lucasfilm. 1977. Movie connected to one of the biggest international media franchises, including numerous games in many genres and across all platforms.
   NOTE FROM THE EDITOR TO THE COPY-EDITOR: I am not sure how to reference this. Suggestions are welcome.
18 DICE. *Mirror's Edge*. Electronic Arts. PC, PS3, and Xbox 360. 2008.
19 Square Enix. *Final Fantasy XIII*. Square Enix. PC, PS3 and Xbox 360. 2009.
20 Kok, Dorus, Boot Yentel, Christiansen, Frederikke & David Dijkhuis. 2020. Video Games as Infrastructures of Resistance. https://mastersofmedia.hum.uva.nl/2020/10/video-games-as-infrastructures-of-resistance/
   Ho, Christopher J. H. 2022. "Civil Disobedience in the Era of Videogames." British Journal of Chinese Studies 12 (2): 101–113. https://doi.org/10.51661/bjocs.v12i2.187.
21 Lilja, Mona. 2022. "The Definition of Resistance." *Journal of Political Power* 15 (2): 202–220. https://doi.org/10.1080/2158379X.2022.2061127.
22 Dalby, Simon. 2017. "Anthropocene Formations: Environmental Security, Geopolitics and Disaster." *Theory, Culture & Society* 34 (2–3): 233–252. https://doi.org/10.1177/0263276415598629.
23 Lilja, 2022.
24 Malm, 2021: 147.
25 Malm, 2021.
26 Malm, 2021.
27 Temper, Leah, Sofia Avila, Daniela Del Bene, Jennifer Gobby, Nicolas Kosoy, Philippe Le Billon, Joan Martinez-Alier, et al. 2020. "Movements Shaping Climate Futures: A Systematic Mapping of Protests against Fossil Fuel and Low-Carbon Energy Projects." *Environmental Research Letters* 15 (12): 123004. https://doi.org/10.1088/1748-9326/abc197.
28 Goldstein, Yoni. 2024. Chicago '68 [Rule book]. Jim Dietz foundation. https://www.1968games.com/.
29 Weiss, Alexandre, Catoire, "Counit" Franck. 2021. Which Side? [Rulebook]. Yeast Games.
30 https://boardgamegeek.com/boardgame/296332/disarm-the-base.
31 https://boardgamegeek.com/boardgame/251141/zone-a-defendre.
32 Flanagan, Mary, and Mikael Jakobsson. 2023. *Playing Oppression: The Legacy of Conquest and Empire in Colonialist Board Games*. Cambridge: The MIT Press. ISBN: 978-0-262-37371-5.
33 Goldstein, 2024.
34 Alexandre, Weiss, Franck, "Counit", Catoire. *Which Side?*. Board game. (2021; Yeast Games).
35 Metheringham, Jessica. 2019. Disarm the Base [Instructions]. Dissent Campaigns and Games Ltd, page: 4.
36 Metheringham 2019, page: 5.
37 Malm 2021, page: 68–69.
38 *Zone à Défendre*. Board game. (Contrevents, 2017).
39 "Zone to Defend, Wikimedia Foundation, last modified 11-10-2024, https://en.wikipedia.org/wiki/Zone_to_Defend.
40 Fremeaux, Isabelle, Jordan, Jay, Herbst, Marc. 2021. We are 'Nature' Defending Itself, Entangling art, activism and autonomous zones. Pluto Press. ISBN 978 0 7453 4587 1, page: 57.
41 https://fr.ulule.com/zadlejeu/.
42 Herbst 2021, page 2–3.

Reflections from a Game Dev Activist

23

Jennifer Estaris

ABSTRACT

This chapter shows how, by allowing ourselves to change and then expressing this change and our new perspectives through games, we can then come together and change the world.

The first part of this chapter describes the author's journey toward working with games and sustainability and toward activism. The second part uses the games the author has designed and worked on and their process as examples for how change and activism can be brought into our design work. This chapter closes by encouraging the reader to join existing groups, work together as activists, and realize that even systemic change is more realistic than it may seem.

KEY TAKEAWAYS

- The personal story and advice from one of the most accomplished and successful game designers, in the space of games for sustainability, and in general
- An honest perspective on the journey that is part of doing this work
- Inspiration and a path to follow for how to bring activist work and systemic change into your own design.

DOI: 10.1201/9781003512400-27

A SEA-CHANGING JOURNEY

Sea-change is a gradual metamorphosis. The term originates from Shakespeare's *The Tempest*, where Ariel sings a song to comfort Ferdinand, explaining his father's body has been slowly changed by the sea into something beautiful and precious: bones into coral, eyes into pearls. Sea-change may be slow, but it's transformative.

I strive for a sea-change in my work and want us all to move toward conscious sea-change. And ideally sea-change with the same transformation, the same destination in mind: what good looks like for the world. I search for a shared vision of a sustainable future. Search with me.

SEA-CHANGING MYSELF

I grew up in the suburbs of Atlanta, Georgia, taking on more than my fair share of environmental harm via mindless consumerism and high-carbon travel modes. A child of Filipino immigrants, I encountered racism and reacted by treading lightly. Isolated, my family turned to games, frequenting arcades with my mom, who loved Ms. Pac-Man. Juxtaposed against the digital life was nature: I lived by a wooded lake and would sit on the roof stargazing amidst chirping crickets.

In high school, I loved two things: words and the world. I was editor of my high school literary magazine and published a tiny 'zine with pen pals from far-off, exotic places: Finland, the Philippines, and Maine. I was drawn to games that explored our relationship with the world; *EcoQuest: The Search for Cetus*[1] taught the importance of environmental ethics and *Below the Root*[2] depicted a mystical and evocative world covered in enormous trees. The player bridges estranged communities and thus prevents world disaster. I modified games like *Lemonade Stand*[3] and the classic card game black-jack, re-balancing them to be less penalizing, more joyful.

In college, I was interested in economic sustainability, obtaining my Bachelor of Science in Economics at the Wharton School, University of Pennsylvania, but I lacked a clear picture of what viable economics should look like. Years later, I attended a camping festival where activist Satish Kumar brought the port into focus: "The economy is a means to an end, and the end is human well-being and planetary well-being. But we don't have that."[4] He underlined how the economy practiced in today's world has distanced itself from its original meaning: *eco*, from the Greek *oikos*, means home or place of residence, and *-nomia* refers to management. Thus, *oikonomia* is about the management and care of our home. We are not caring for our home due to our violent exploitation of resources and uber-capitalist behaviors. Instead, modern economics has been reduced to optimizing money. Kumar recommended we rename "economy" to "moneynomy."

Midway through moneynomy school, I realized I needed the arts to create meaning. I needed to care for the soul, which I'll call *animanomia*. Thus, I added on English literature, which snowballed into a MFA in Creative Writing at Columbia University, which then swerved when I took a game design course during grad school, where Professor Bernard Yee (*Stay: Forever Home*) introduced me to the alluring and mass-reaching world of game development. If I wanted to connect with people, with other souls, and form a positive future, true global *animanomia,* what better way than through interactivity?

Games I've worked on have had cozy environmental themes. At Nickelodeon in the early 2000s, a *Go! Diego Go!* arcade game[5] allowed players to guide a sea turtle to jellyfish or plastic bags. In-person playtesting revealed that some children enjoyed the sea turtle's cough reaction when eating plastic. Treading that line between realistic teachings and entertaining fantasy was tough. Easier were subtle normalizations, like ensuring *Neopets' Petpet Park,*[6] a virtual world where alien pets played together, had vegetarian food.

In addition to making environmental games, I connected with people who showed *Why* and *How* to take a larger stand on climate action. The "Why" was thanks to my child, as I worried about the future, I would be leaving for her. I worried about the future I would be leaving to billions of children around the world. How much of my undoing could I undo?

The "How" was shown by those like Mathias Gredal Nørvig, CEO of SYBO Games, known for the astoundingly mass-reaching *Subway Surfers*[7] (4 billion downloads![8]). Mathias lived and breathed sustainable values while cultivating a commercially successful studio. He gave talks on sustainability in games. A sea-changing question emerged: Could I also take a stand?

I started slow, with small talks in safe spaces. Now, I speak regularly at conferences like GDC and SXSW on the intersection of games and social change: *How One Billion Gamers Can Save the World*[9] or *Cozy Futures*: *Challenging Traditional Game Design.*[10] More recently, the talks have included a game performance, sampling out my tabletop role-playing game *Terra Preta*. These sessions allowed me to engage with sustainability luminaries like Deborah Mensah-Bonsu, Trista Patterson, and Sam Barratt. Sam, the Chief of Youth, Education and Advocacy at the United Nations Environment Programme, highlighted the importance of a climate community. No surprise, as Sam, alongside Mathias, is a co-founder of Playing for the Planet, a UN-facilitated alliance with over 50 game studios collaborating on green initiatives. I am proud to consider myself a part of this caring community. And all I had to do was pipe up. Activism can be approachable.

My efforts on climate action are not just limited to conferences. I take to the streets. The Restore Nature Now march in London[11] ended with a die-in where 50,000 protestors lay on streets. It was surprisingly cozy: the warmth of the asphalt below, the blue skies behind tree crown shyness, and an odd, comforting silence while snuggled together on the road in various wildlife costumes. There was something enticing and memorable about this gentle, cozy setup. Activism can be cozy.

Activism can be playful. The global environmental movement Extinction Rebellion frequently has family-friendly events, and I host a booth with nature-related board games. One event had us sprawled out in the marine animals' wing of the Natural History Museum playing Ocean Bingo.

The climate movement is iterating. It's more playful. More approachable. More cozy. Just like games.

SEA-CHANGING GAMES

Popular activism games of the past tended toward action-packed eco-terrorism, such as the heart-wrenching, unrivalled *Final Fantasy VII*.[12] Nowadays, cozy activism games are coming to a head, like *I Was a Teenage Exocolonist*,[13] *Beecarbonize*,[14] and the Financial Times' *The Climate Game*.[15] These experiences normalize activism and spotlight its impact.

I joined ustwo games to direct *Monument Valley 3*[16] and wanted to explore cozy activism.

ustwo games is a B-Corp certified game studio committed to positive impact. For example, *Alba: A Wildlife Adventure*[17] shows how even the smallest person can make a difference, and ustwo games has planted over 1.5 million trees with the support of Alba's players. *Assemble with Care*[18] is a game about repairing technology, extending the life of objects.

Let's zoom into ustwo's most known games, the *Monument Valley* series. Here, players manipulate Escher-inspired architecture through an enigmatic, beautiful world. The first *Monument Valley*[19] launched in 2014 and was a sleeper hit, showing that small screens can have a big heart and memorable art. The series has over 150 million downloads, with cultural staying power, evidenced by its presence in museum exhibits and pop culture, like Ariana Grande's music video *No Tears Left to Cry*.[20]

The Green Game Jam: Monument Valley 2's The Lost Forest

Transforming a beloved world like *Monument Valley* proved challenging. My team and I took the sea-change approach, starting with something small. We created a DLC for *Monument Valley 2*,[21] as part of Playing for the Planet's Green Game Jam (GGJ).

The GGJ is an annual project organized by Playing for the Planet where gaming studios are challenged to implement "green activations" into their games for positive real-world impact. For example, *Subway Surfers* had a Bali event in a rainforest setting, with wind turbines in the background, and sold special items for charity.

The GGJ kicks off with expert talks. One speaker, Peter Wohlleben,[22] author of *The Hidden Life of Trees*,[23] gave his talk in a lush, dark forest. His nonfiction book surprisingly reads like a fairy tale, about how trees communicate through the hormones that spread in the wind, through fungal networks that connect their roots. He explained how trees not only protect their children as shelter from the wind but also strengthen them by shading them, giving them less sun, so they become more resilient trees.

We wanted to do something inspired by Wohlleben's talk, thinking about trees, shrouded in a beautiful darkness, caring for each other, sharing their stories. We started with research: Monument Valley's cutting room floor, nature games like *Below the Root*, and most importantly, trees. Senior Artists Fen Beatty and Jessie van Aelst spoke of the Japanese tree-pruning technique *daisugi*, which reminded CEO Maria Sayans of a tree she encountered on her daily walks, which reminded me of Wohlleben's description of cadaver rejuvenation.

Monument Valley is a love letter to world architecture. For the GGJ, could we create a love letter to forests? The team created scenes where architecture intertwines with trees and darkness, which formed a chapter in Monument Valley 2. In this chapter, titled The Lost Forest, players explore impossible architecture hidden in nature in order to reconnect with the beauty of trees. This chapter reminds players of our emotional connection to trees, so that we remember to protect them… which then protects us.

When the Monument Valley 3 team joined the GGJ, many of the team members had just finished a game that was all about environmental activism, *Alba: A Wildlife Adventure*. The game innocently begins as a bird watching game, and eventually Alba takes down the corrupt hotelier trying to destroy the local nature reserve. One of Alba's actions involves running around the community collecting petition signatures.

The *Lost Forest* chapter begins and ends with a call to action: to sign the Play4Forests petition, which calls on world leaders to protect forests. It's an emotional, poetic appeal, with a throwback to *Alba*. The petition also asked players questions about their interests in environmental content in games. This Playing for the Planet survey was also offered by nine other GGJ participants' games, resulting in almost 400k responses. The results[24] are promising:

- 81% would like to see more environmental content in their games if it's relevant to the experience
- 61% are motivated to pay for environmental content
- 73% pledged to take action to change their behavior

By the time Monument Valley 2's *The Lost Forest* launched in October 2021, the core game was over five years old. The downloadable content (DLC) caused a resurgence of interest in the game, resulting in app store features, new reviews, and higher rankings, all of which translated to more sales.

By being open to inspiration from activities like the GGJ and cultivating green conversations, beauty can blossom. Beauty, action, reactivation, and yes, money. Money does grow on trees.

After the GGJ: Monument Valley 3

After the GGJ, our team focused on developing *Monument Valley 3*. We wanted to continue giving our players hope, as we did in *The Lost Forest*, while also evolving the 3rd installment.

I want to ask Banksy, Miyazaki, the Guerilla Girls, and other artivists how they translated a need for a deep change in the world's value system into their work. For me, systems change begins from within. My yearning for a sustainable future was as unstoppable as a rushing river. As the director, I distilled key values into approachable visions: *Eco-villages. Regenerative balance. Hopepunk.* My teammates and I blended these visions with art, tech, and gameplay, and unraveled their combination into pillars. We tried many things: multiplayer, user-generated content, a town builder, first person. These didn't work for various reasons.

So we jammed. We split into multi-disciplinary exploration pods for a few weeks, aiming for ludonarrative cohesion. One pod explored sailing, referencing limited sailing experiments attached to splines. In that pod, Manesh Mistry untethered us from those constraints, referencing an open-world sailing prototype he jammed on a while back. I recalled my years in idyllic Sluseholmen, a canal town in Denmark while I was working at SYBO. I kayaked my child to her school and freely jumped from my balcony into the clean canal waters. I remembered my mother's childhood stories in the Philippines, and how her father's fishing boat was saved by an eel. These memories were all joyful and gentle, with a tinge of surrealism. For sea-change, what was our port? In *Sky: Children of the Light*,[25] the developers wanted players to enjoy the freedom of flying. For sailing in *Monument Valley 3*, we wanted to capture the gentle joy of sailing.

The pod created a magical sailing prototype that fit the game's pillars. The game-feel was tightened, and the storyline organically emerged. Noor, a lighthouse keeper's apprentice, must find a new source of light before the world drowns in darkness. She searches by sailing, giving a feeling of open-world, contained.

We collaborated closely with climate experts, both Netflix's sustainability team and our own in-house experts. Christa Avampato, a writer and biomimicry scientist getting her Master's in Sustainability Leadership at the University of Cambridge, joined us during our concept phase. Christa encouraged us to tell a story that's personal so players could imagine how they might make these choices in real life. During pre-production, Trevin York was brought on to support aspects of climate resilience and adaptation as a climate designer and senior fellow at the Atlantic Council's Climate Resilience Center, alongside Dr. Clayton Whittle. Trevin recommends embedding a climate expert on a team. Generally, the expert can highlight design strategies for game elements to showcase the team's chosen climate themes and share inspiring related resources. We leveraged their strengths based on our needs for that phase; for example, Christa was part of a narrative pod, and Trevin was part of a social impact system design pod.

Their climate expertise laid the foundations for our social impact partnership with the International Federation of Red Cross and Red Crescent Societies (IFRC) to support communities affected by devastating floods. Through this collaboration, we raise awareness of flood risks and contribute directly to relief efforts, helping rebuild lives and foster resilience. Funds help deliver emergency shelter and supplies, medical care, and community recovery such as infrastructure for sustainable futures. A huge nod to Maria Sayans, Jane Campbell, and Jamie Wotton for enabling this partnership.

Playtesting and iteration is an essential element of honing the Monument Valley craft. We leaned upon the climate community, especially the IGDA Climate SIG. The community gifted us with playtesting and answered a Google form survey consisting of quantitative and qualitative answers. Dedicated players, like superfan Parker Wallace, shared further via a PlaytestCloud video recording. Playtesting helped us strike the balance of being true to the Monument Valley universe while exploring new waters.

Monument Valley is typically a solitary experience. In contrast, *Monument Valley 3* has many characters who abandon their destructive ways after a catastrophic flood.

Eventually, they form a regenerative, nature-based community. Meghna Jayanth's essay *White Protagonism and Imperial Pleasures in Game Design*[26] re-energized us: "We should as designers be in collusion with our players, defectors to the dominant powers who benefit from our collective numbness, inaction, atomisation and exploitation."

We gently push for collective action by depicting collective action.

In *Monument Valley 3*, yes… followed by the real world?

SEA-CHANGING THE WORLD

In Spring 2024, Playing for the Planet organized an international Earth Week of Action. In London, I hosted a "Lost Forest" litter-picking walk, complemented by plant identification from foraging expert Elena Hoge (*Out and About*[27]) and a tree-bath meditation from Aster & Goldenrod's co-founder Valentina Marchetti. Gwen Boissel ran a Climate Fresk. Louisa Keight and Angel Spence-Lewis organized a clothing swap. Tristan Clark (*Love & Pies*[28]) and Nic Walker hosted a pub night to relax and reboot with game devs and climate activists.

But we need more people in the movement. In the words of renowned climate justice activist Farhana Yamin, "I'm hoping that people tap into that courage that's needed when there's a true emergency and release the heroes in themselves."[29] The 3.5% rule shows that a small, engaged minority can change the world. Harvard's Carr Centre for Human Rights Policy found that nearly all non-violent protests succeed once 3.5% of the population participates.[30] In the UK, that's 2.4M[31] people; in my hometown Atlanta, 16,235[32]; in the game industry, 11,550.[33] We can do this.

As Gandhi said, "If you want to change the world, start with yourself." Join initiatives like the IGDA Climate SIG or Playing for the Planet. Share your personal journey or game, and they'll gladly share climate expertise and general feedback. Start small, start with what you have, and iterate bit by bit. Sea-change. Distill those visions and then collaborate to create pillars.

By embracing personal transformation, we influence those around us, creating ripples of change that grow into waves. Climate action isn't an isolated effort: it thrives within communities, which are crucial to countering activism fatigue. Together, we amplify our individual contributions and inspire one another, like winds guiding each other to the port we need.

Finally, be gentle. Creating the world we want—a gentle, sustainable future—must be created in a gentle, sustainable way. Listen to others, to the world, to yourself, to your soul. *Animanomia*. Then, after rest and reflection, take to the vast sea, follow the winds, and transform again.

NOTES

1   Sierra Entertainment. *EcoQuest: The Search for Cetus.* Sierra Entertainment. PC. 1991.
2   DeSharone, Dale. *Below the Root.* Windham Classics. Apple II. 1984.

3   Jamison, Bob. *Lemonade Stand*. Minnesota Education Computing Consortium. Apple II. 1973.

4   Satish Kumar, "Emergence of New Paradigm" (Bluedot Festival, Cheshire, England, July 23, 2022).

5   Nick Jr. Games. *Go Diego Go!: Tuga the Sea Turtle*. Nickelodeon. Web. 2005.

6   Neopets. *Petpet Park*. Nickelodeon. Web. 2008.

7   SYBO. *Subway Surfers*. SYBO. Mobile. 2012.

8   Jeffrey, Rousseau, "Subway Surfers surpasses 4bn downloads." *GamesIndustry.biz*. March 20, 2023. https://www.gamesindustry.biz/subway-surfers-surpasses-4bn-downloads-news-in-brief.

9   Jennifer Estaris, "How One Billion Gamers Can Save the World" (South by Southwest, Austin, Texas, March 17, 2022).

10 Jennifer Estaris, "Cozy Futures: Challenging Traditional Game Design" (devcom Developer Conference, Cologne, Germany, August 20, 2024).

11 "The Wildlife Trusts, "Over 60,000 people march to parliament to demand politicians Restore Nature Now," 23 June 2024, https://www.wildlifetrusts.org/news/over-60000-people-march-parliament-demand-politicians-restore-nature-now. Jim Powell, "Restore Nature Now march in London – in pictures," *The Guardian*, June 22 2024, https://www.theguardian.com/environment/gallery/2024/jun/22/restore-nature-now-march-in-london-in-pictures. Extinction Rebellion, "Restore Nature Now! Activists and environmental groups come together for largest march for nature ever seen", June 21 2024, https://extinctionrebellion.uk/2024/06/21/restore-nature-now-activists-and-environmental-groups-come-together-for-largest-march-for-nature-ever-seen/.

12 Square. *Final Fantasy XVII*. Square Enix, Square, Sony Computer Entertainment, and Eidos Interactive. PC, PS, PS4, Android, Nintendo Switch, and Xbox One. 1997.

13 Northway Games. *I Was a Teenage Exocolonist*. Finji. PC. 2022.

14 Charles Games. *Beecarbonize*. Charles Games/Steam. PC. 2023.

15 Financial Times. *The Climate Game*. Financial Times. Web. 2022.

16 ustwo games. *Monument Valley 3*. Netflix Games. Mobile. 2024.

17 ustwo games. *Alba: A Wildlife Adventure*. ustwo games. Mobile. 2020.

18 ustwo games. *Assemble with Care*. ustwo games. Mobile. 2029.

19 ustwo games. *Monument Valley*. ustwo games. Mobile. 2014.

20 Grande, Ariana. *No Tears Left to Cry*. Music Video. 2018.

21 ustwo games. *Monument Valley 2*. ustwo games. Mobile. 2017.

22 Wohlleben, Peter. Green Game Jam Keynote Speech. Playing for the Planet. March 2021.

23 Wohlleben, Peter. The Hidden Life of Trees: What They Feel, How They Communicate: Discoveries from a Secret World. Greystone Books. 2016.

24 Ower, Jude, and Mathias Gredal Nørvig. 2024. Gaming for Good. Rethink Press, 122–123.

25 Thatgamecompany. *Sky: Children of the Light*. Thatgamecompany. Mobile. 2019.

26 Jayanth, Meghna. White Protagonism and Imperial Pleasures in Game Design #DIGRA21. Web. 2021. https://medium.com/@betterthemask/white-protagonism-and-imperial-pleasures-in-game-design-digra21-a4bdb3f5583c.

27 Yaldi Games. *Out and About*. Steam. PC. Forthcoming.

28 Trailmix. *Love & Pies*. Trailmix. Mobile. 2020.

29 Yamin, Farhana. Quoted in interview "Farhana Yamin: 'It took 20 minutes to unglue me from Shell's office. It was a bit painful'" by Andrew Anthony. *The Guardian*. May 19, 2019.

30 Erica Chenoweth. "The success of nonviolent civil resistance" TEDxBoulder. November 4, 2013.

31 Worldometer. https://www.worldometers.info/.

32 Worldometer. https://www.worldometers.info/.

33 Uni Global Union. https://uniglobalunion.org/report/the-video-game-industry.

Take Care of Each Other

24

Cultural and Personal Frameworks for Surviving the Game Industry

Hexe Fey

ABSTRACT

This chapter offers the personal reflections of the author on how to survive in the game industry while trying to have a positive impact. It considers more broadly the reasons and goals for working with games and then discusses the three main elements the author considers central: Building a community for mutual support, interacting with the game industry as if it is in collapse, and setting up support for and prioritizing mental health, this chapter also features several strategies and pragmatic recommendations for getting these three elements into place.

KEY TAKEAWAYS

- The author's personal approach to surviving in the game industry
- The importance of communities of mutual support and how to find them
- Guidelines for interacting with the game industry as if it as in collapse
- Advice on how to set up support and care for one's mental health

DOI: 10.1201/9781003512400-28

When I started making games, it was out of joy, with little hope or care if I would ever be paid. As I got to know the game industry better, it became apparent that I would definitely never be paid. Or at least, not paid in the same way other people were paid. I was never going to be able to support myself with the money from games. Years later, that continues to be true, but along the way, I developed a sort of system, a framework to inform how to build my projects and live a somewhat comfortable life.

It isn't a workflow—it is more like a worldview and ethic that helps me choose what is a priority, what is a healthy working environment, and how to set boundaries when things are not as good as they should be. I believe our bodies and our spirits are our first and ultimately only sustainable resources. If we do not take care of ourselves, we cannot build beautiful things. As someone who has worked freelance gigs for most of my adult life, sorting out how to set boundaries and balance work with other work with living is a necessity.

How would one design a lifestyle that allows for interacting with both the current game industry, the requirements of capitalism to feed, clothe, and house yourself while also preserving your own health and sanity? How would one do that while also constructing this system to be flexible enough to extend into an increasingly unstable future? Within the topics of sustainability and games and industry are more existencial questions nested. We need to consider how we, mammals with thumbs and billions of microorganisms making our electric brain meats more or less conscious, sell our labor and create objects that may or may not be art or may or may not be products. A bigger puzzle is: why? Why would we spend our limited time making these things? Is it for love? Is it for money? The absurdity of the whole of modern human civilization looms above us while we grapple clumsily with the existentialism of labor and pixels and sustainability. Ultimately, nobody can answer these questions for anyone but themselves, but I can explain how I organize my own small universe.

As a person with what some have labeled a "complicated" background, I find myself writing this, which is definitely not a manifesto, to encourage people to take care of themselves and their environments. My background as a community harm reductionist informs my work and perspectives as much as my mixed Lakota and European-American heritage. Specifically, these intersecting backgrounds highlight the necessity of being aware of the reach and repercussions of one's actions.

For example, Will doing a project with a company that has abusive labor practices cause harm to myself and others? Will using different types of computing and tools require the use of so much electricity and water that it significantly contributes to climate collapse? Will networking and socializing with people whose highest goal is making money through the suffering of others cause psychic damage? How many jobs and money-making methods are worth it? How does my relationship with my environment, my community, and myself inform my interactions with capitalism? This unfortunately always leads to the question: How long can I go without needing money? The different dimensions of sustainability work and ecological approaches that are represented throughout the book, for example in this chapter from David and Thorsten et al., are represented here as well.

Ignoring money for the time being and the complex budgeting that a conscious avoidance of laboring for capitalists requires, there are three main elements that I have found helpful.

First is a community network, both professional and personal. Having professional networks is invaluable for getting and vetting jobs. Having personal networks of people connected to culture, whether that culture is indigenous, art and aesthetic based, labor unions, academic, book clubs, potlucks, or even subculture based doesn't matter. Connection to other humans is what matters and it is what will save a person from houselessness and hunger.

My networks of collaborators and community have helped me eat, live, travel the world, and exposed me to art and games I never would have found on my own. Building a healthy community network is no small task and often requires navigating personalities that might not mesh well with your own and identifying inequitable power dynamics. Communication and transparency about why and what people want and need are useful tools.

Some markers of healthy community networks I have observed are a lack of ego-driven personality cults, a culture of self-education and curiosity, active support of vulnerable and less privileged people through access to professional opportunities without gatekeeping, support of basic physical needs and emotional needs without shaming, respecting elders and children, and a habit of double-checking rumors and news. Community accountability and response to harm must be mentioned as well, but it is such a huge, complex topic that I cannot do it justice here other than to say that the previous markers of healthy communities contribute significantly to the ability to do accountability and respond to harm.

Second is interacting with the game industry, and by a less obvious extension, the arts economy as if it is in collapse. This might seem unreasonably doom-oriented, but the current formation of the game industry's ever-expanding profits, the arts economy of galleries and rich donors, and the academia around games and art has proved to be unstable at best. Like most capitalist-based systems, it will have to radically change its business models in order to survive. What that looks like, I cannot predict, but it is wise to plan for companies and organizations to collapse, for layoffs and consolidation of workers to be common, and for uncertainty to be the norm. Backup plans and flexibility are important.[1]

Regarding backup plans and flexibility, professional and community networks facilitate resources and connections for the backup plans, the business changes, and the resource pooling, union organizing, collective bargaining, and if necessary, legal processes that come with precarious industries. Also, returning briefly to the question of money: when budgeting for precarious incomes, I have found it useful to avoid high-cost habits. What that means to you is up to your interpretation and relies heavily on the cost of living where you are and what is a necessary extravagance, but doing an analysis of what and where you spend money can help.

Of course, personal lifestyle changes do not change the machinery of industry and capitalism. It can only help us navigate it and living like a monk or hermit can only be fulfilling for most (but not all) people for a short period of time before the need to live fully overtakes us. Joy is a human necessity, and in a culture based on exchanging money, it unfortunately sometimes requires having some.

Third is mental health support. Mental health and associated support systems don't have the same stigma it did for the generation before, but it is still complicated and difficult to navigate. This can be particularly true if you are a person with intersectional needs. For me, culturally relevant mental health support is the key. Psychology is based on culture, and having a counselor or therapist approaching you from a perspective that conflicts with your own cultural values is worse than useless—it is harmful. Setting up mental health support before a crisis or a burnout is highly recommended, and taking time to rest, pacing your work projects, and if possible, setting reasonable goalposts for development timelines is also highly recommended.

Humor, joy, and connection with the world outside of work are also important parts of supporting your own mental wellbeing, but outside of setting an alarm to tell you to go touch grass (which has varying levels of success) achieving perspective and balance can be difficult. Some suggestions I have found useful: laughter (any kind), talking to a friend, reading a novel, considering the connection of waters and rocks all over the world and how every piece of earth touches another piece of earth that touches your home and also every place you have ever felt at home or marveled at the beauty and power of nature, and also snacks.

These tools and frameworks are in alignment with my own ethics and cultural values, which, out of necessity, negates some of the individualism that makes up the core of Western cultural tendencies. An intellectual and cultural relative of this has been outlined in *How To Build Anything Ethically*[2] by Suzanne Kite (in conversation with fellows) in the Indigenous AI Position Paper. While Kite's paper uses Lakota cultural ethical frameworks for developing technology and offers ways to relate to and build hardware and physical machines, and ultimately, anything, the piece you are currently reading encourages applying mutualistic ethics and planning to the way you interact with labor and yourself.

The concepts that connect this paper and Kite's are culturally and ethically relevant to myself and other indigenous peoples, and from my observations also to many non-indigenous peoples. Culturally appropriate systems of ethics regarding research and development of technology vary greatly across the globe, and as this piece focuses on sustainable ways to navigate the game industry it is also attempting an impossible goal: to be useful regardless of cultural background or affiliations. I encourage everyone to research, adapt, and create their own culturally appropriate systems.

If your goal is to get rich and gain fame through exploitation and unethical extraction, this framework I have laid out here is not for you, and I am surprised you have made it through this essay at all. If your goal is to work in the game industry any way you can possibly do it, regardless of the ethics of your employers, I recommend at least having mental health support in place. If your only options are to work for the megacorps and AAA sweatshops, try to get what you can from them and take care of your humanity.

However, you are probably a fellow traveler, in which case you are welcome, and I hope we can all live in a good way together.

NOTES

1 Depending on where you live in this instability is more or less apparent and extreme. I live in the USA as of writing this paper.
2 Kite, Suzanne, "*How to Build Anything Ethically*", Indigenous Protocol and Artificial Intelligence Position Paper, (2020): 75, https://www.indigenous-ai.net/position-paper/.

Conclusion

25

Patrick Prax, Clayton Whittle, and Trevin York

This book aimed to empower you by offering a vision for meaningful climate action, a concrete starting point no matter how you're involved with game making, and a pathway to connect the two. We have shown the approaches of people who are working at the cutting edge of this topic in the design and content of their games, in the way they are changing game making, and in their ways of engaging as human beings. The theory of change for this connection stresses the importance of building community, working together, and supporting each other—because it is clear that the fight against climate change will be a long process of re-organizing society over decades. It is a process that we need to accelerate in earnest, but also one that will require support and a community.

Thank you for staying with us throughout this book. Now this is the place where we have shared with you the most important lessons that we had learned so far. Now we are looking forward to learning from you and to engaging in this struggle together. To return to the metaphor at the start: we've shown you the path to the dragon. You're geared up and ready to go with us towards that challenge that will, on some level, decide over the future of this world. We will still, together and with the help of each other, need to build our community, learn to read and understand the terrain for the final battle, hone our skills, and develop our tactics. The work continues.

So please join us in the IGDA Climate SIG and in the green game movement more broadly. Everyone is needed, and all perspectives can contribute. Share your unique approach in your game, your studio, or your advocacy group, so that we can develop and grow together. We are looking forward to learning from you, and everything that works is valuable.

If you are a teacher, then we are looking forward to further developing good pedagogical approaches and practices together in the SIG or in other forums. Hearing from you how using this book in teaching functions in your context would be especially valuable.

DOI: 10.1201/9781003512400-29

If you are a student looking to join the game industry or to make games, then please know that there is a community here that cares deeply about sustainability and is excited to welcome you and your contribution.

We extend our heartfelt thanks to all our authors, readers, peers, and caring human beings out there!

Patrick, Clayton, and Trevin

Index

For Product Safety Concerns and Information please contact our EU
representative GPSR@taylorandfrancis.com
Taylor & Francis Verlag GmbH, Kaufingerstraße 24, 80331 München, Germany

9 7 8 1 0 3 2 8 3 6 1 9 5